# Construction Health and
# Safety Management

# Construction Health and Safety Management

**Alan Griffith**
**Tim Howarth**

**PEARSON**
Longman

Harlow, England • London • New York • Boston • San Francisco • Toronto
Sydney • Tokyo • Singapore • Hong Kong • Seoul • Taipei • New Delhi
Cape Town • Madrid • Mexico City • Amsterdam • Munich • Paris • Milan

**Pearson Education Limited**
Edinburgh Gate
Harlow
Essex CM20 2JE
England

and Associated Companies throughout the world

*Visit us on the World Wide Web at:*
http://www.pearsoned.co.uk

First published 2001

ISBN 0 582 41442 3

**British Library Cataloging-in-Publication Data**

A catalogue entry for this title is available from the British Library.

10  9  8  7  6  5
07  06  05  04  03

Set by 35 in 9/13 Palatino
Printed in Malaysia

# Contents

# List of figures

# List of tables

# List of abbreviations

| | |
|---|---|
| APAU | Accident Prevention Advisory Unit |
| BS | British Standard |
| BSI | British Standards Institution |
| CDM | The Construction (Design and Management) Regulations 1994 |
| CHSWR | The Construction (Health, Safety and Welfare) Regulations 1996 |
| CIRIA | Construction Industry Research and Information Association |
| COSHH | The Control of Substances Hazardous to Health Regulations 1996 |
| DETR | Department of the Environment, Transport and the Regions |
| EC | European Community |
| ECI | European Construction Industry |
| EEC | European Economic Community |
| EMS | Environmental Management System |
| EU | European Union |
| HSC | Health and Safety Commission |
| HSE | Health and Safety Executive |
| H&SMSA | Health and Safety Management System Assessment |
| HSWA | The Health and Safety at Work, etc. Act 1974 |
| IiP | Investors in People |
| IMS | Integrated Management System |
| IMSA | Integrated Management System Assessment |
| ISO | International Standards Organization |
| LFS | Labour Force Survey |
| MHSWR | The Management of Health and Safety at Work Regulations |
| PPE | The Personal Protective Equipment at Work Regulations 1992 |
| QA | Quality assurance |
| QMS | Quality Management System |
| RIBA | Royal Institute of British Architects |
| RIDDOR | The Reporting of Injuries, Diseases and Dangerous Occurrences Regulations 1995 |
| SHE | Safety, Health and Environment |
| TQM | Total quality management |

# Preface

The management of health and safety is without doubt the most important function of construction management. Construction work is inherently dangerous. Injuries to persons on and around construction sites occur regularly. It is fortuitous that many injuries are minor, but, others are serious and some are fatal. The construction industry has over recent decades suffered a poor safety record. While the number of fatal accidents demonstrated a welcome decline in the 1990s this should not encourage complacency. Construction management has a perpetual and unswerving challenge to ensure a safe and healthy working environment.

The Construction (Design and Management) Regulations 1994, known as the CDM Regulations, introduced welcome and much needed legislation. It augments an existing framework of legislation for health and safety at work and brings a clearer vision for the management of construction safety, health and welfare. The CDM Regulations are concerned with the management of health and safety throughout the total construction process. Clear responsibility is placed upon clients, designers and contractors to be proactive in the planning, coordination and management of health and safety. The Regulations focus on identifying the potential hazards to health and dangers to safety during each stage of the construction process together with the assessment of their risk.

The CDM Regulations require the delivery of a project health and safety plan. Divided into two parts, the first stage focuses on the project evaluation and design processes with the objective of producing a pre-tender health and safety plan. The second part focuses on the production site processes with the objective to produce a construction phase health and safety plan. It is the essential element of planning which forms the basis for the management approach with risk assessment as its core theme.

Effective health and safety management systems and working procedures are the goal of the main parties to a construction project. The client's planning supervisor is charged with delivering the pre-tender health and safety plan, while the principal contractor is charged with delivering the construction health and safety plan. Moreover, the principal contractor must establish management systems and working procedures which ensure the maintenance of safe working conditions in and around the project site. Well formulated health and safety management systems will identify, assess and control risk both within and across the professional boundaries of the various contractual parties. Feedback loops within the designer's and contractor's approach will ensure that information is directed not only within the span of control of the individual party but contributes to the management processes within other systems. Under the CDM Regulations, effective management will be evidenced in the production of the health and safety file – a complete profile of health and safety planning and management throughout the construction project. The CDM Regulations place a clear responsibility upon the principal contractor to develop, implement and maintain effective health and safety management procedures on its project sites.

This book focuses on the principal contractor. It is suggested in this book that the implementation of a well considered and conceived, formally structured and well organized health and safety management system is an appropriate way for a principal contractor to meet its responsibilities, not only within the scope of the CDM Regulations but also in meeting other current legislation. Such a system, supporting both the corporate and project-based organizations, will meet the requirements of BS 8800, the specification for health and safety management systems. Health and safety management systems (H&SMS) development is following much in the same way as quality management and environmental management with the rapid evolution of accredited certification schemes. There is little doubt that the health and safety management system will become an established and prominent aspect of construction management in the early part of the new millenium.

# Acknowledgements

The authors gratefully acknowledge *all* those persons who contributed to the production of this book.

# Part A  The nature of construction health and safety

# 1 Introduction

## The hazardous nature of construction

Poor health and safety management costs the construction industry millions of pounds each year. Across the UK economy the cost to business and commerce runs into billions of pounds annually. Yet, more fundamental and of greater importance, poor health and safety management costs lives. Construction continues to be one of the most hazardous industries within which to work. The Egan report *Rethinking Construction* (DETR, 1998) stated that 'the health and safety record of construction is the second worst of any industry' and suggested that 'accidents can account for 3–6 per cent of total project costs'. Injuries, accidents and dangerous situations are commonplace. Over the last 20 years the construction industry has suffered a poor health and safety record. While the number of fatal accidents declined throughout the 1990s this must not encourage any sense of comfort and complacency. 'The rates of death, serious injury and ill health are still too high' (HSE, 1997). The industry has a perpetual and unswerving challenge to deliver and maintain a safe working environment. Effective health and safety management demands a clear vision, a systematic approach and a sustained commitment to improvement.

## Health and safety: a paramount consideration

Health and safety is a paramount consideration for all construction industry professionals. It impinges upon the work of the planning authorities, clients, consultants, contractors – in fact anyone who works on or is in the vicinity of a construction project. Whereas in the past health and safety management has traditionally been perceived as a production site activity, it has now become a holistic consideration and integrated responsibility of duty holders across the total construction process. It has also become an integral part of the corporate organizational framework and structure as well as the operational management of many organizations who contribute to the construction process.

## Awareness for construction health and safety

The level of awareness and recognition for health and safety within the construction industry is increasing. Advances in health and safety management practices have been rapid following the implementation of the Construction (Design and Management) Regulations 1994, or CDM Regulations, (HSE, 1994), which came into effect in 1995. The CDM Regulations augment much additional legislation including The Health and Safety at Work, etc. Act 1974 (HSE, 1974), commonly referred to as the HSWA, The Management of Health and Safety at Work Regulations 1992 (HSE, 1992) and The Construction (Health, Safety and Welfare) Regulations 1996 (HSE, 1996). The CDM Regulations have made a profound contribution to the cultural shift in safety, health and welfare legislation which has evolved over the last 30 years.

## The CDM Regulations

The CDM Regulations apportion clearly the duties and responsibilities of clients, designers and contractors together with those of new duty holders – the planning supervisor and principal contractor. CDM also requires new documentation – the 'health and safety plan' and the 'health and safety file' – to be developed and implemented. It is clear, however, that the level of experience of construction professionals across the industry is currently limited. One would expect this and there is much learning to be absorbed before everyone becomes totally comfortable with the application of CDM.

While responsibilities for health and safety under CDM are shared more than they have been in the past, a weight of responsibility falls on the principal, or main, contractor. Many principal contracting organizations will realize that the concept and principles of good health and safety management are the same as effective construction project management. Moreover, they will recognize that successful project health and safety is supported by an effective corporate organization which shows commitment to strong health and safety policies, procedures and practices.

## Health and safety management systems

The key driver to achieving a safe and healthy working environment is to ensure that health and safety issues are assessed, planned, organized, controlled, monitored, recorded, audited and reviewed in a systematic way. An appropriate way for the principal contractor to address the legislative requirements, corporate business needs and practical project demands of health and safety is to establish a health and safety management system within the organization. Such an approach can be a dedicated system or one which forms part of an existing organizational management system, such as a quality system. Through the establishment of a formal management approach to health and safety, the principal contracting organization will be well equipped to develop a strong and positive health and safety culture. It will be able to develop the policies, plans, management procedures and safe working practices essential to achieving successful construction health and safety management. Moreover, it will add value to its organizational culture, enhancing its ability to maintain its core business and to attract new business in the future.

## Structure of the book

This book is structured in three parts. Part A presents a comprehensive introduction to the nature and demands of construction health and safety, Part B presents the framework of health and safety legislation and Part C focuses on the development and application of effective health and safety management by the principal contractor. It does not intend to be prescriptive but to present key considerations. Compliance with current legislation is quite specific but the methods by which compliance is achieved is a matter for each principal contractor to determine. Construction health and safety management is, put simply, a paramount responsibility facing the principal contractor today and one that will increase in importance in the future as legislation becomes ever more stringent, customer perception magnifies and industry expectations increase.

## References

Department of the Environment, Transport and the Regions (DETR) (1998) *Rethinking Construction*, Construction Task Force, HMSO, London.

Health and Safety Executive (HSE) (1974) *The Health and Safety at Work, etc. Act 1974*, HMSO, London.

Health and Safety Executive (HSE) (1992) *The Management of Health and Safety at Work Regulations 1992*, HMSO, London.

Health and Safety Executive (HSE) (1994) *The Construction (Design and Management) Regulations 1994*, HMSO, London.

Health and Safety Executive (HSE) (1996) *The Construction (Health, Safety and Welfare) Regulations 1996*, HMSO, London.

Health and Safety Executive (HSE) (1997) *Health and Safety in Construction 1997*, HMSO, London.

# 2 Blackspot Construction

## Introduction

This chapter presents an overview of the Health and Safety Executive report *Blackspot Construction* (HSE, 1988). The report presents the findings of a study of five years' fatal accidents in the building and civil engineering industries. Although the study was undertaken in the early to mid 1980s it is *the* benchmark document for health and safety management in construction. It provides an essential foundation for understanding the nature of construction health and safety.

## *Blackspot Construction*

*Blackspot Construction* (HSE, 1988) is considered to be the definitive report on the occurrence of accidents within the construction industry. The report, based on the detailed findings of a unique research study, analyses the circumstances leading to the loss of life of 739 persons within the building and civil engineering industries between 1981 and 1985. The overriding claim of the report is that these lives could have been saved by the better management of health and safety within the construction industry.

*Blackspot Construction* highlights unequivocally that the principal reasons for fatal accidents within the construction process result from fundamental lapses in attention to health and safety. These are: a general lack of foresight to danger; the absence of supervision; insufficient education and training for identifying and meeting potential hazards; and, a general lack of attention to detail.

The reasons underlying these tragedies should not in any way be rationalized by explanations founded in the difficulties and complexities of construction. It is well accepted that construction is overshadowed by a unique combination of circumstances: its temporary work site; the transient workforce; and, busy and often over-stressed management. The report clearly highlights that accidents in construction are avoidable. It should be clearly recognized that they are avoidable. Moreover, the construction process should be managed with an optimum awareness for, and the highest regard for, health and safety so that fatal accidents are inherently avoided.

The message in the 1988 report, from the Director General of the Health and Safety Executive, J. D. Rimmington, emphatically conveys the responsibility that the construction industry must bear with respect to health and safety:

> Construction work needs to be organized in such a way that it does not continue unnecessarily to claim the lives of so many, including fit and experienced workers.

To all that plan, manage and supervise construction work he suggests:

> Ask yourself; Could this happen in my company? Could this happen to one of our contractors? Could this happen to me? . . . and make sure it never does!

## The principal accident statistics

**The facts**  The Health and Safety Executive determined that in the five-year period 1981–85, 739 accidents within the construction industry resulted in fatality. Of these, 561 were accidents involving employees and 120 involved self-employed workers, 94 of which were working as subcontractors and 26 working for themselves. Furthermore, 37 accidents involved members of the public and 21 involved children.

Table 2.1 presents fatal accident statistics for the five-year period. It can be seen clearly that the annual number of fatal accidents to construction employees had remained almost constant in each year ranging from 109 in 1981 to 113 in 1985. Greater variability can be seen in the other groups recorded. In the category of self-employed subcontract workers the number of fatalities doubled from 12 to 23 during the period, while the number of fatalities of children almost tripled from 3 to 8.

**Table 2.1** Number of fatal accidents in construction industry, 1981–85

| Group | 1981 | 1982 | 1983 | 1984 | 1985 | Total |
|---|---|---|---|---|---|---|
| Employees | 109 | 112 | 119 | 108 | 113 | 561 |
| Self-employed working as subcontractor | 12 | 18 | 23 | 18 | 23 | 94 |
| Self-employed working on their own | 7 | 5 | 8 | 4 | 2 | 26 |
| Members of the public | 9 | 9 | 8 | 5 | 6 | 37 |
| Children | 3 | 4 | 4 | 2 | 8 | 21 |
| Total | 140 | 148 | 162 | 137 | 152 | 739 |

*Source*: HSE (1988)

**Table 2.2** Number of fatal accidents in construction recorded by occupation, 1981–85

| Occupation | Employees |
|---|---|
| Labourers and civil engineering operatives | 193 |
| Roofing workers | 68 |
| Painters | 40 |
| Drivers | 47 |
| Demolition workers | 43 |
| Managerial and professional status | 29 |
| Carpenters and joiners | 30 |
| Scaffolders | 22 |
| Steel erectors | 20 |
| Bricklayers | 15 |
| Plumbers and glaziers | 10 |
| Electricians | 7 |
| Other construction and non-construction occupations | 37 |
| Total | 561 |

*Source*: HSE (1988)

Table 2.2 presents the distribution of the 561 fatalities to employees among occupations within the construction. Labourers and civil engineering operatives comprise the largest group, accounting for 34 per cent of all fatal accidents. Trade operatives that typically work at heights, roofing workers, painters, scaffolders and steel erectors were involved in 68, 40, 22 and 20 fatal accidents respectively. Combined these account for 27 per cent of all incidents. A further occupational group that suffered substantial fatalities was that of other construction and non-construction workers. These are occupations that typically service and support the construction processes on site. Some 37, or 7 per cent, of total fatal accidents occurred in this group.

Almost three-quarters (74 per cent) of all accidents occurred within the building sector. Incidents involved construction and demolition works and occurred during maintenance operations to houses, commercial and industrial processes. The remaining 26 per cent of total accidents involved works to the construction and maintenance of roads, sewer pipelines, sea and harbour defences and large petrochemical, oil and gas installations within the civil engineering sector. These statistics are shown in Figure 2.1.

Figure 2.2 presents the main categories of fatal accidents within the construction sector as a whole for the five-year period. Clearly, falls represent the principal reason for accidents, accounting for 52 per cent of the total number occurring. Accidents involving falling materials or objects and transport and mobile plant account for 19 per cent and 18 per cent respectively. Electrical hazards, asphyxiation (which includes drowning), fire and explosions and

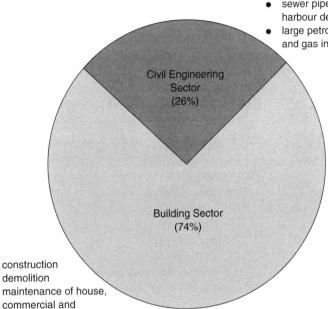

**Blackspot Construction**

- construction and maintenance of roads
- sewer pipelines and harbour defences
- large petrochemical, oil and gas installations

Civil Engineering Sector (26%)

Building Sector (74%)

- construction
- demolition
- maintenance of house, commercial and industrial premises

**Fig. 2.1** Distribution of fatalities by construction sector *Source*: data from HSE (1988)

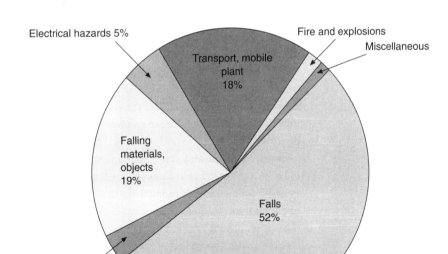

Electrical hazards 5%

Fire and explosions

Miscellaneous

Transport, mobile plant 18%

Falling materials, objects 19%

Falls 52%

Asphyxiation, drowning 3%

**Fig. 2.2** Major categories of fatal accidents in construction *Source*: HSE (1988)

miscellaneous reasons account for the remaining 11 per cent of the total fatal accidents within the construction processes.

**What do these principal accident statistics show?**

The principal accident statistics demonstrate that incidents are not exclusive to any occupational group nor to specific activities. Accidents occur across the total construction process and to any participants, even those who are time-served and well experienced. Often, fatal accidents occur during simple and routine works during activities which might normally be considered to be quite innocuous, for example painting operations. Whereas a high level of risk might reasonably be expected, for example in demolition, maintenance work presents a surprisingly high risk with a substantial number of fatal accidents occurring.

The HSE propounds that:

> Experience is no safeguard and experienced workers were just as likely to be killed as trainees.

This reinforces the need for training, in particular for operatives new to the inherent dangers of construction. Equally, education and training remains essential to experienced workers as the HSE points out:

> People are often at risk because recognized custom and practice in the industry is not necessarily safe practice.

Construction operatives succumb to quite unnecessary accidents. Falling from ladders, working platforms and structures, falling through roofs, the collapse of excavations, the impact from falling materials and dangers from plant movement are the most prominent. Accidents can occur in any situation, whether a construction project be small or large, confined or extensive, complex or simple. Accidents can be just as likely to occur from a major and transparent danger, such as the inappropriate deployment of a mobile crane, as they can from poor housekeeping around the site, for example not tidying up debris on a working platform. The HSE suggests that such instances were inexcusably marred by a lack of planning, inadequate supervision and unsafe systems of work.

Effective health and safety management starts with pre-site planning and the development of safe systems of working for any construction project. This must be followed with proactive attention to health and safety awareness by operatives and staff and with positive leadership and attentive supervision by managers. Health and safety management requires liaison between all those involved with the project – the client, consultants, principal contractor, subcontractors, professional advisers, project staff and workforce, and the public.

Everyone involved with a construction project and working on or visiting a construction site must learn to identify danger and take positive steps to protect themselves and others.

## The analysis of fatal accidents within construction

This section presents an analysis and evaluation of the fatal accidents examined by HSE and reported in *Blackspot Construction* (HSE, 1988).

**Falls**  Over half, 52 per cent (383 out of 739), of all fatal accidents recorded in the HSE study resulted from falls. This is the largest single cause of death among construction workers.

Figure 2.3 presents an analysis of the category of falls broken down into groups, of which there are six.

- Falls from roofs – 37 per cent (142).
- Falls from scaffolds and working places – 31 per cent (121).
- Falls from ladders – 17 per cent (66).
- Falls during demolition and dismantling works – 8 per cent (29).
- Falls during steel erection works – 3 per cent (11).
- Falls due to other causes, for example during erection of cranes, and hoists, and falls resulting in drowning – 4 per cent (14).

### Falls from roofs

Three main situations accounted for the 142 fatal accidents resulting from falls from roofs. These were: (1) roof edge falls; (2) falls through fragile materials; and (3) falls from the internal structure of roofs.

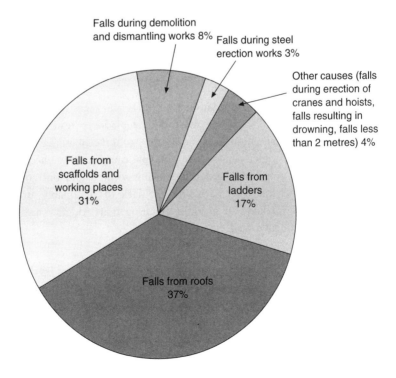

**Fig. 2.3** Breakdown of fatal accidents resulting from falls *Source*: adapted from HSE (1988)

The most prominent causes of roof edge falls were the poor condition of roof ladders and the lack of proper securing to the structure, workers slipping down roofs because roofing ladders were not used, and the absence of edge protection. Falls through fragile materials were associated with two main causes, these being that fragile roofs were not covered with crawling boards, and openings, such as roof lights, were not covered. Falls from the internal structure of roofs occurred principally from walking on sheeted panels between the supporting frame members.

### Falls from scaffolds and working places

The 121 fatal accidents which occurred in this group are attributed to the following situations: (1) falls during the erection and dismantling of scaffolds and temporary works; (2) falls through openings in working platforms; (3) falls from tower scaffolds; (4) falls from cradles, bosun's chairs and suspended scaffolds; and (5) falls from scaffolds and working places not attributed to the aforementioned situations.

The main causes of fatal accidents during erection and dismantling of scaffolding were that platforms did not have edge protection, platforms slipped or collapsed, and that platforms were not provided. Accidents during temporary works found their causes in inappropriate formwork and falsework. Falls from tower scaffolds involved situations of the scaffold collapsing or over-turning and persons falling from the working platform. Falls from cradles, bosun's chairs and suspended scaffolds were due mainly to defective installation, poor maintenance, use of defective materials in the equipment and poor supervision and training of operatives. The causes of falls in other situations involving scaffolding were a lack of guard rails and toeboards and the untidiness of working platforms.

### Falls from ladders

Falls from ladders and stepladders resulted in 66 fatal accidents. These situations involved an operative working alone or working with one other person. The principal causes of these accidents were rooted in ladders and stepladders not being secured and not correctly footed during use at the workplace and the lack of reasonably practicable precautions being taken, for example ensuring that ladders were placed at the correct angle to a wall.

### Falls during demolition and dismantling works

Eight per cent of fatal accidents from falls involved situations which occurred during demolition and dismantling works. The cause of all of these accidents was falling from working places. In all cases, poor housekeeping, such as an untidy work platform and failure to take reasonably practicable precautions while undertaking tasks were in evidence.

### Falls during steel erection works

Three per cent of fatal accidents from falls were attributed to situations occurring during steel erection operations. These were traced to failures in taking reasonably practicable precautions such as working without a safety harness.

### Falls due to other causes

Four per cent of fatal accidents from falls occurred from a number of disparate situations. These were falls during the erection of cranes and hoists, falls resulting in drowning, falls from workplaces less than two metres above the ground and falls at ground level. In all these situations accidents occurred from not taking reasonably practicable safety precautions and a poor standard of site housekeeping.

**Falling materials and objects**

The second largest category of fatal accidents was denoted falling materials and objects, representing 19 per cent of the total (143 out of 739).

Figure 2.4 presents a breakdown of this category of fatal accidents by group. There are three groups as follows:

- Collapse of structures or parts of structures – 41 per cent (59).
- Insecure loads, unsecured equipment and pieces of plant – 31 per cent (44).
- Falls of rock or earth from sides of excavations and tunnels – 28 per cent (40).

All of the fatal accidents within this category occurred when materials, objects or earth fell on operatives.

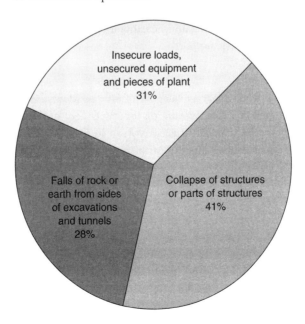

**Fig. 2.4** Breakdown of fatal accidents resulting from falling materials and objects *Source*: adapted from HSE (1988)

### Collapse of structures or parts of structures

The 59 fatal accident situations within this group involved: falling materials and objects from demolition and dismantling works; the collapse of walls; the collapse of structures being erected; burial under collapsing floors; and collapsing chimneys.

Many of these accidents occurred during demolition works where the collapse of buildings or structures resulted in operatives falling or being crushed. In almost all cases the cause was a lack of adherence to safe working practices. The causes of wall collapse were cited as being: design faults; inability to withstand pressure from debris and back-filling; undermining by neighbouring exacavations; and poor ground and rain affected conditions.

### Insecure loads, unsecured equipment and pieces of plant

The 44 fatal accidents which occurred within this group involved being struck by falling materials and loads, and by pieces of equipment and plant. These accidents involved not only construction operatives but members of the public, including children.

The main causes of these accidents were: poor practices during loading and unloading; poor maintenance of equipment; poor slinging of materials; and the inadequate delineation of the work site.

### Falls of rock or earth from the sides of excavations and tunnels

Forty fatal accidents occurred during excavation operations and within excavations. Typical excavations were for water mains, sewers and drainage systems. Most situations occurred in ground conditions thought to be safe and also in shallow works where collapse was thought unlikely.

The main causes of excavation collapse were: absence of earthwork support; inadequate or failing earthwork support; and, unsafe practices in installing earthwork support.

**Transport and mobile plant**

The third largest category of fatal accidents involved the use of transport and mobile plant, accounting for 18 per cent of the total (137 out of 739).

The type of vehicle involved in these fatal accidents was significant, with heavy plant and bulk transport vehicles involved in most situations. In rank order the types of vehicles involved in transport accidents were:

- Excavators, shovels and earth moving equipment.
- Dumpers and dump trucks.
- Lorries.
- Vans, private cars, buses, and motorcycles.
- Mobile cranes (including lorry mounted).
- Forklift trucks.
- Others (road rollers, mobile working platforms).

The type of accident situations occurring from transport and mobile plant were, in the following rank order groups:

- Run over or struck by vehicle moving forward – 49 per cent.
- Run over or struck by vehicle reversing – 18 per cent.
- Falls – 11 per cent.
- Falls of equipment or materials – 10 per cent.
- Vehicles overturning or stationary – 3 per cent.
- Others – 9 per cent.

The main causes of these accidents were: unsafe systems of work and transportation; failures to provide a safe place of work; poor training, instruction and supervision; and, unsafe acts by operatives, for example riding on plant not designed to carry passengers.

**Electrical hazards**
Electrical hazards accounted for 5 per cent of the total number of fatal accidents (34 out of 739).

Three main groups form the breakdown of this category of fatal accident, as shown in Figure 2.5. The groups are:

- Unearthed or live equipment – 73 per cent (25).
- Contact with overhead power lines – 18 per cent (6).
- Contact with underground cables – 9 per cent (3).

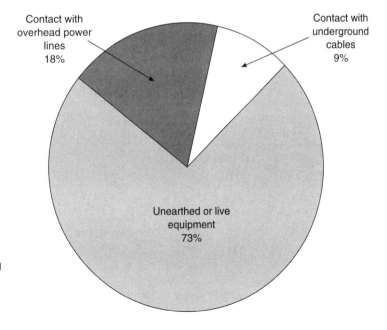

**Fig. 2.5** Breakdown of fatal accidents resulting from electrical hazards *Source*: adapted from HSE (1988)

The main causes of accidents from electrical hazards were: equipment touching live power lines; lack of safety instruction; absence of safety signs and barriers; working without location plans for live cables; and unsafe excavation practices when working near cable routes.

**Other accidents**     Other accidents account for 6 per cent of the total number of fatal accidents (42 out of 739).

This category is broken down into three sub-categories as shown in Figure 2.6. These sub-categories are:

- Asphyxiation which includes drowning – 55 per cent (23).
- Fires and explosions – 29 per cent (12).
- Miscellaneous – 16 per cent (7).

**Asphyxiation and drowning**

The main fatal accident situations recorded in this sub-category were: gassing by carbon monoxide and fumes from fuel gas heaters; lack of oxygen when working in confined spaces; and drowning when working on coastal and river works.

The causes of these accidents were: lack of safe working systems; poor instruction and supervision; absence of emergency procedures; and inappropriate use of equipment.

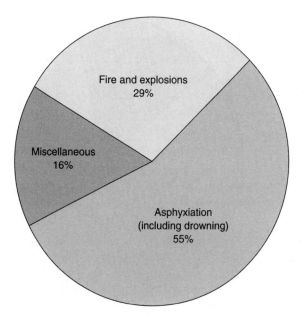

**Fig. 2.6** Breakdown of other fatal accidents within construction *Source*: adapted from HSE (1988)

### Fires and explosions

The fatal accident situations attributed to this sub-category were: fires in site hutting and caravans; falls into fires during disposal of rubbish; burns through the use of faulty equipment; contact with debris from blasting operations; and gas explosion through damaging domestic service pipes.

The causes of these accidents were traced to the lack of safe working practices and poor instruction and supervision.

### Miscellaneous

As a percentage of the number of fatal accidents occurring in the category 'Other accidents', 7 out of 42, or 16 per cent, were sub-categorized as miscellaneous. No details were provided for this sub-category.

**Maintenance works**  Maintenance work can be highlighted as a particularly high-risk activity within construction. *Blackspot Construction* identifies that over the five-year review period 43 per cent of all fatal accidents occurred during maintenance works (315 out of 739) as shown in Table 2.3.

Figure 2.7 presents a breakdown of fatal accidents occurring during maintenance works. There are five main groups within this category:

- Roofing works – 41 per cent.
- Falls from scaffolds – 19 per cent.
- Falls from ladders – 18 per cent.
- Transport – 13 per cent.
- Electrical works – 9 per cent.

Within each of these groups the situations giving rise to fatal accidents are the same as those outlined previously under the main category headings. Similarly, the causes underlying these fatal accidents were as outlined previously.

**Table 2.3** Number of fatal accidents occurring during maintenance works, 1981–85

|  | *1981* | *1982* | *1983* | *1984* | *1985* | *Total* |
|---|---|---|---|---|---|---|
| Number of maintenance accidents | 57 | 74 | 72 | 60 | 52 | 315 |
| Total number of construction accidents | 140 | 148 | 162 | 137 | 152 | 739 |
| Maintenance accidents expressed as a % of the total number | 41% | 50% | 44% | 44% | 34% | 43% |

*Source*: HSE (1988)

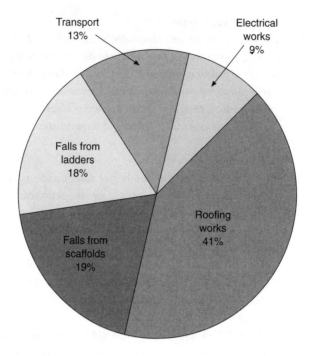

**Fig. 2.7** Breakdown
of fatal accidents
occurring during
maintenance works
*Source*: adapted from
HSE (1988)

The main point to note is that maintenance work is dangerous. Often these works are carried out in what might seem quite innocuous situations. In fact, maintenance is frequently carried out in unsafe buildings, confined spaces, at height, or in occupied premises and therefore maintenance work intrinsically contains all the components that can lead to hazardous working.

**The public and children**

Of the total number of fatal accidents occurring within construction 37 members of the public and 21 children were killed.

The majority of these accidents occurred outside of normal working hours, mainly in the evenings and weekends. However, some accidents did occur while the site was occupied and children are particularly vulnerable as they tend to find construction sites exciting and sometimes an inviting place to play.

The main types of fatal accidents to children involved: plant and transport; falling materials and equipment; and falls from heights and through openings.

The main causes of these accidents were; poor site security; inadequate warning signs; failure to cover exposed works, accesses and openings; and failure to secure and immobilize plant and equipment.

## The key causes of construction accidents

Throughout *Blackspot Construction* (HSE, 1988) a great many accident situations were identified and range of causes determined. In addition to fatal accidents, it is well accepted that several thousand accidents resulting in serious and minor injury are recorded each year and also that many accidents are simply not reported (HSE, 1994).

If the causes of all accidents occurring within construction are analysed three main determinants emerge:

1. There was a clear lack of liaison and coordination between construction professionals at the pre-construction stage leading to inadequate planning for safe working practices.
2. Practices then current within the construction industry were not encouraging project participants to identify hazardous situations and implement sensible safe working practices.
3. There was a lack of instruction and supervision by managers to ensure that safe working practices were implemented and maintained.

It is worth noting that, at that time, non-traditional forms of contractual arrangement, greater use of subcontracting and increased use of self-employed labour meant that remote-management was sometimes being employed. This brought problems for project management in ensuring that principal contractors and subcontractors were using safe systems of work.

It was this background of practice within the construction industry, typified by the findings of *Blackspot Construction*, that led to the radical and rapid development of comprehensive and stringent legislation in the late 1980s and early 1990s. This resulted in the implementation of European Directives and their associated national Regulations of European Community (EC) member states. Such legislation seeks to enforce the development, implementation and maintenance of safe systems and practices of working and more effective health and safety management within the construction industry.

## Key points

This chapter has identified that:

- *Blackspot Construction*, the definitive report by the Health and Safety Executive (HSE, 1988), presents the findings of a study of five years' fatal accidents within the construction industry.
- In the five-year study period 739 persons lost their lives needlessly as a result of construction accidents.

- The overriding claim of the report was that loss of life could have been avoided by the better management of health and safety.
- 'People are often at risk because recognized custom and practice in the industry is not necessarily safe practice' (HSE, 1988).
- Everyone involved with a construction project must learn to identify danger and take steps to protect themselves and others.
- Effective health and safety management starts with pre-site planning and the development of safe systems of working.
- Successful health and safety management will be based on sound corporate and project systems which considers health and safety as a major contributor to organizational success.

## References

Health and Safety Executive (HSE) (1988) *Blackspot Construction: A study of five years fatal accidents in the building and civil engineering industries*, HMSO, London.

Health and Safety Executive (HSE) (1994) *CDM Regulations: How the Regulations Affect You*, HMSO, London.

# 3  Recent health and safety statistics

## Introduction

The Health and Safety Executive (HSE) gathers detailed, accurate and reliable information on workplace injuries, across a range of industries. Notwithstanding, only a proportion of injuries, accidents and hazardous incidents are reported under statutory legislation. Within construction this proportion is only around 40 per cent. There is little doubt that the discipline of reporting injuries, accidents and hazardous incidents needs to be improved. This chapter outlines how information on accidents is gathered and how this reveals the extent of reporting, level of risk, and accident trends within the construction industry.

## Collection of health and safety statistics in the United Kingdom

The HSE conducts an annual Labour Force Survey (LFS) to gather accurate and reliable information on workplace injuries across a range of industry sectors. This information complements the data reported under statutory health and safety legislation, for example The Health and Safety at Work, etc. Act (HSE, 1974) and The Reporting of Injuries, Diseases and Dangerous Occurrences Regulations (RIDDOR) 1995 (HSE, 1995).

**Levels of reporting**   It is well recognized in almost all industries, and within the construction industry in particular, that only a small proportion of injuries, accidents and hazardous incidents are reported under statutory legislation. This can be seen clearly if the number of injuries reported to enforcing authorities is compared with the number of injuries recorded in the Labour Force Survey conducted by the HSE. An example of this can be seen in the 1995–96 data which, to provide more accurate figures, combines with data for the preceding and succeeding years. The number of injuries reported in the LFS was 380,000 whereas the number reported to enforcing authorities was 150,000. The level of reporting injuries under statutory legislation was therefore only 40 per cent. This is exacerbated when it is remembered that far from all injuries, accidents

**Table 3.1** Level of non-fatal injury in construction relative to other industries

| Industry | Level of risk |
| --- | --- |
| Mining | 2.66 |
| Construction | 1.51 |
| Transport, storage and communications | 1.10 |
| Agriculture | 1.02 |
| Manufacturing | 1.00 |
| Health and social work | 0.86 |
| Public administration and defence | 0.85 |
| Consumer/leisure services | 0.73 |
| Distribution, repair, hotels and restaurants | 0.66 |
| Business | 0.33 |
| Education | 0.33 |
| All industries | 0.78 |

*Source*: adapted from HSE (1996)

and hazardous incidents will be admitted to the Labour Force Survey. Many health and safety incidents are simply never reported.

Within construction, the HSE suggest that in 1995–96 the reporting level was 30 per cent (HSC, 1996). For self-employed persons, and there are many within the construction industry, the reporting level is suggested by HSE to be less than 10 per cent. From the 11 industry sector types recorded by HSE construction is listed sixth in the ranked order for levels of reporting of occupational injuries. The topmost ranked industries were energy and water industries, public administration (government), and transport and communications. Within construction the discipline of reporting injuries, accidents and hazardous incidents needs to be improved.

**Levels of risk**   The data gathered by the LFS and RIDDOR give a good indication of the level of risk to which a person is exposed within their respective industry. If 'manufacturing' is used as a benchmark, and for comparative purposes assigned a value of 1.0, other industries can be ranked according to their comparative level of risk. Table 3.1 shows 11 industry sectors ranked in order of level of risk. It can be seen that data for 1995–96 (HSC, 1996) lists construction as having the second highest level of associated risk behind the mining industry. A risk value of 1.51 indicates that the risk of injury to an employee in construction is over 50 per cent higher than in manufacturing. Construction remains, therefore, an intrinsically hazardous industry.

**Accident trends**   There have been welcome reductions in the annual number of fatal and non-fatal injuries in almost all sectors of industry over the last decade. These range

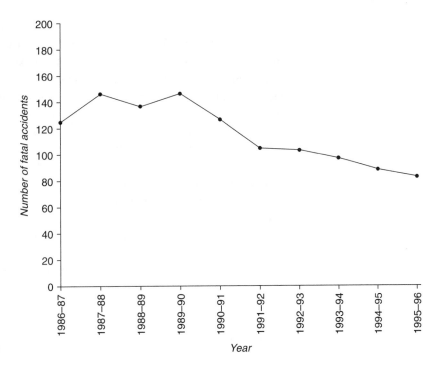

**Fig. 3.1** Profile of fatal accidents within construction between 1987 and 1996
*Source*: HSE statistics

from a 9 per cent reduction to accidents in public administration occupations to a 48 per cent reduction in the energy and water supply industries.

Within construction the profile of fatal accidents between 1987 and 1996 is shown in Figure 3.1. The profile indicates clearly that following the harrowing statistics recognized in *Blackspot Construction* (HSE, 1988) and a peak rate in annual fatal accidents in 1989–90, there was a declining rate throughout the 1990s. A 30 per cent reduction in fatal accidents within construction was witnessed between the late 1980s and the late 1990s. This trend in falling numbers of annual fatal accidents is surely welcomed by all who work in and are associated with the construction industry. Notwithstanding, it should not generate complacency and the challenge to industry to reduce accidents further remains.

## International perspective

Health and safety matters impinge upon construction whenever and wherever it is carried out. Construction is a global business. Health and safety is of international interest, generally where organizations operate within their own domestic markets and specifically where organizations transcend national business frontiers.

While the UK construction industry enjoyed improved health and safety performance throughout the 1990s the situation in other countries has been somewhat different. It is not pertinent to even attempt direct comparisons of industrial health and safety across countries. There will be considerable differences in, for example, economic climate, market forces, political environment, construction methods, and availability of resources. Any or all of these factors could make comparison quite erroneous. Notwithstanding, common threads in the broad reasons of why injuries, accidents and hazardous incidents occur within construction are apparent and of interest.

In the UK construction industry three persons in every 1,000 are involved in an accident (HSC, 1996). Research studies conducted in Hong Kong (Lingard and Rowlinson, 1998) highlighted that in 1991 a peak figure of 374 persons per 1,000 workers suffered an accident while working within construction. Lingard and Rowlinson make the comparison that while nearly one-third of all construction workers in Hong Kong were involved in an accident this was more than twice the rate of construction accidents in the USA (150 per 1,000), 25 times that of Japan (15 per 1,000) and 30 times that of Singapore (12 per 1,000).

## A view on 'health' and safety

Much research and many publications, including this book, focus on construction safety. Construction health and welfare is, however, frequently given less attention. It is often asked if construction is a 'safe place' to work, but, rarely is it asked if construction is a 'healthy place' to work.

Construction takes place in temporary work situations which are influenced greatly by the particular type of work being carried out. Also, there are physical changes taking place continually throughout the duration of the project which means that conditions may be healthy and safe one minute and distinctly hazardous the next.

Operatives on site are exposed or have the potential to be exposed to a wide range of physical, chemical, biological, mechanical, and psychosocial influences and difficult environmental working conditions. Weather, temperature change, confined spaces and working at heights can all take their toll. While one can safeguard workers to some degree, these and other factors, acting singularly or in combination, can have very serious negative effects upon them.

Occupational exposures to adverse agents are well researched and the damage they cause are regularly and comprehensively reported by the HSE. Many diseases, illnesses and injuries can be caused within construction occupations, as indeed they can within almost all industry occupations. In addition, there can be a whole host of related problems which can occur, for example, alcohol abuse, stress and physical and mental side effects. *The ECI (European Construction Institute) Guide to Managing Health in Construction* (Gibb *et al.*, 1999) presents comprehensive facts and figures for occupational health matters.

It is asserted that 'health' should be considered equally as important as 'safety'. Therefore, an effective health and safety management subsystem must embrace this within its management plans, procedures and working instructions. This is imperative given the ever present hazards and danger to people which is intrinsic to construction works.

## Key points

This chapter has identified that:

- Only a proportion of injuries, accidents and hazardous incidents are reported under statutory legislation and many health and safety incidents which occur within construction are simply never reported.
- The construction industry is recognized as having a high level of occupational risk and, relative to the principal industries categorized by the HSE, construction is the second most hazardous industry in which to work.
- There was a 30 per cent reduction in the recorded number of fatal accidents occurring within construction between the late 1980s and the late 1990s.
- Health and safety matters impinge upon construction whenever and wherever it is carried out and so are of international interest. It is recognized that some countries have an appallingly high annual number of health and safety incidents and therefore an effective health and safety management approach is of global interest.
- An effective health and safety management system must recognize that 'health' is equally as important as 'safety' and must embrace this within its management plans, procedures and working instructions.

## References

Gibb, A. G. F., Gyi, D. E. and Thompson, T. (1999) *The ECI (European Construction Institute) Guide to Managing Health in Construction*, Thomas Telford Publishing, London.

Health and Safety Commission (HSC) (1996) *Health and Safety Statistics 1995/96*, Government Statistical Service, HMSO, London.

Health and Safety Executive (HSE) (1974) *The Health and Safety at Work, etc. Act 1974*, HMSO, London.

Health and Safety Executive (HSE) (1988) *Blackspot Construction: A study of five years fatal accidents in the building and civil engineering industries*, HMSO, London.

Health and Safety Executive (HSE) (1995) *The Reporting of Injuries, Diseases and Dangerous Occurrences Regulations (RIDDOR) 1995*, HMSO, London.

Lingard, H. and Rowlinson, S. (1998) 'Behaviour Modification: A New Approach to Improving Site Safety in Hong Kong' *Construction Management and Economics*, Vol. 16, No. 4, pp. 481–488.

Health and Safety Commission (HSC) (2000) *Health and Safety Statistics 1999/2000*, London.

# 4 The cost of construction accidents

## Introduction

This chapter focuses on the losses caused by construction accidents. Accidents cost the construction industry millions of pounds annually. Such is the magnitude of the problem that a definitive cost has never been determined. Useful studies have been carried out by the Accident Prevention Advisory Unit (APAU) (HSE, 1993a) in an attempt to identify the full cost of accidents. These have raised awareness among organizations in many industry sectors of the cost implications of accidents and this may serve in the future to stimulate better health and safety practices. What is certain is that the cost of accidents can be characterized by what the HSE terms the 'accident iceberg', see Figure 4.1 on p. 29. While the tip of the iceberg reveals visible and tangible costs, the submerged bulk of the iceberg harbours a great many hidden and often indeterminate financial implications. This chapter provides the basis for understanding the fundamental aspects and issues associated with the cost of accidents including those which occur within construction.

## The cost of industry failures in health and safety

The Accident Prevention Advisory Unit of the Health and Safety Executive suggests that 'the cost of failures in health and safety management are high' (HSE, 1993b):

- 30 million days lost in a year from work-related injuries and ill health (approximately £700 million per annum).
- A two-thirds increase in real terms of employers' liability insurance costs over the past decade and a doubling of claims since 1985.
- Uninsured losses from accidents (whether they result in personal injury or not) can cost anything between 8 and 36 times what an organization normally insures for; in some cases making the difference between profit and loss.

Detailed case studies undertaken by APAU (HSE, 1993a) were aimed at developing a methodology to accurately identify the full cost of accidents. Organizations from various different industries were involved in the studies.

The results from the studies suggested that the accidents cost:

- one organization as much as 37 per cent of its annualized profits;
- another the equivalent of 8.5 per cent of (the project's) tender price;
- a third organization 5 per cent of its (annual) running costs.

Irrespective of how accident costs might be quantified the simple fact is that accidents cost the UK economy billions of pounds each year (HSE, 1993a). The costs to a single organization can also be considerable and in a worst case scenario may make the difference between a company continuing to trade or going out of business. In addition, and of paramount importance, there is the cost in human terms and one cannot and should not put a price on a life.

## Construction health and safety costs

When considering the costs and benefits of implementing the Construction (Design and Management) Regulations 1994 (HSE, 1994) the Health and Safety Commission (HSC) determined that the total cost to the construction industry would be approximately £550 million.

> The Commission recognized that the reduction in the level of accidents would be the principal quantifiable benefit. They assumed that on small to medium-sized sites, the reductions in accidents would be 33 per cent if the Regulations were implemented, whereas on large sites, where safety management is usually better developed, a 20 per cent reduction in accidents could be expected. The Commission concluded that the estimated benefit to the industry would be £220 million each year.
>
> (Joyce, 1995)

Again, it is evident that the cost of accidents, even when seen as a potential saving to the construction industry, is colossal, running into millions of pounds annually. It should be remembered that these costs are, again, not the full cost of construction accidents. There are many indirect and hidden costs resulting from accidents which are very difficult to quantify. This is particularly true within the construction industry. It is important to recognize that while some costs are easy to see, others are hidden and difficult to put a value to.

## The visible and hidden costs of accidents

The full costs of any accident may be separated into two categories. There are those costs which are visible or 'insured costs', so called because these are the

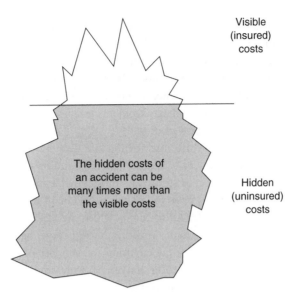

Visible
(insured)
costs

Hidden
(uninsured)
costs

The hidden costs of
an accident can be
many times more than
the visible costs

**Fig. 4.1** The
'accident iceberg'
*Source*: adapted from
HSE (1993a)

costs for which an organization is normally insured. Then there are hidden
or 'uninsured costs' which often far exceed the costs which are insured. The
two categories of accident costs with their disproportionate sizes have been
likened to the principal characteristic of an iceberg; while the small tip of the
iceberg is visible, most of its volume is hidden below the waterline and is
therefore difficult to quantify. Determining the proportion of insured costs to
uninsured costs is extremely difficult. The hidden costs of an accident can be
many times more than the visible costs, as depicted in Figure 4.1.

Visible (insured) costs to a construction organization can include:

- employer's liability
- public (third party) liability
- damage to property.

Hidden (uninsured) costs to both the construction project and corporate
organization can include:

- investigation costs
- legal costs
- fines
- loss of reputation and image
- loss of business and goodwill
- sick pay
- loss due to damage of materials, plant and equipment
- replacement and repairs
- production schedule disruption and delays
- additional resourcing.

In the case studies undertaken by the HSE (HSE, 1993a) five different fields of employment were examined. These were a construction site, a creamery, a transport company, an oil platform, and a hospital. It was stated that 'the participating organizations displayed average, or better than average, health and safety performance in their industries. It is likely that other organizations with less developed management systems would incur larger losses than those identified in these studies'.

Focusing on the construction site element, the main contractor, a wholly owned subsidiary of an international building and civil engineering company, was the subject of study. The project had a value of £8m and a duration of 13 months. It was observed over a period of 18 weeks and involved 120 persons working on site. All accidents which met the definition 'an unplanned event that resulted in injury or ill health of people, or damage or loss to property, plant, materials, or the environment or a loss of business opportunity', met a financial value of £5 or above and were considered by the contractor to be 'preventable' were recorded.

In the study period there were no major injuries (those resulting in over three days absence of the person involved). There were 56 minor injuries (those which required first aid treatment) and 3,570 non-injury accidents. These incidents were estimated to have cost the construction project a direct financial loss of over £87k. Opportunity costs (those incurred where labour was paid but without production) added over £157k to bring the total to over £245k. Over the duration of the contract the estimated total losses were suggested to be £700k, or approximately 8.5 per cent of the tender price. The ratio of insured costs to uninsured costs incurred by the main contractor was 1:11. It was determined by the study that all of these accidents and the costs incurred could have been avoided. Of the accidents that occurred, 832 (23 per cent) were adjudged to have been caused by inadequate planning. Alone these amounted to over £41k (17 per cent) of the total accident costs.

The main theme of this chapter has been to illustrate just how difficult it is to accurately reflect the true cost of any construction accident. Also that, in the main, most accidents are avoidable, particularly with care and forethought given to those activities that take place within the construction process. The principal theme developed in Part C of this book is to incorporate appropriate care and forethought within an effective health and safety management system. Such a system will not only reduce the cost of accidents and injury but meet health and safety legislation, contribute significantly to accident prevention and, moreover, save lives.

## Key points

This chapter has identified that:

- Accidents cost the UK economy billions of pounds each year.
- While it is virtually impossible to put a total cost on accidents occurring within the construction industry it is known that the cost runs into millions of pounds per annum.
- The cost of any accident comprises two main components: 'visible' or 'insured' costs and the much greater proportion of 'hidden' or 'uninsured' costs.
- Irrespective of how accident costs are presented or interpreted it is paramount that one cannot and should not put a price on a life.
- Research studies conducted by the Accident Prevention Advisory Unit (APAU) of the Health and Safety Executive reported that, in their experience, most accidents were avoidable.

## References

Health and Safety Executive (HSE) (1993a) *The Costs of Accidents at Work*, HMSO, London.

Health and Safety Executive (HSE) (1993b) *Successful Health and Safety Management*, HMSO, London.

Health and Safety Executive (HSE) (1994) *The Construction (Design and Management Regulations 1994)*, HMSO, London.

Joyce, R. (1995) *The CDM Regulations Explained*, Thomas Telford Publications, London.

# Part B The framework for health and safety legislation

# 5 The EC legislative framework for health and safety at work

## Introduction

This chapter outlines the role and influence of the European Union (EU) in the development of UK health and safety legislation. It provides an insight into:

- The evolution and function of the European Union.
- The types of legislation of the European Union.
- The institutions of the European Union and their contribution to the legislative process.
- Aspects of Treaties and Acts of the Union relevant to the development of health and safety legislation.
- The Framework Directive – on measures to encourage improvements in the safety and health of workers at work.

The 'further reading' section at the end of the chapter (p. 44) lists publications which focus on EU law and EU law in relation to the construction industry in particular.

## Development of the European Community

The European Union came into existence in November 1993 and serves as an umbrella organization for a number of previously constituted organizations. The constituent organizations of the European Union have served to integrate member countries in an effort to safeguard peace and promote economic and social progress. The constituent organizations are as follows:

- The European Coal and Steel Community (ECSC).
- The European Economic Community (EEC).
- The European Atomic Energy Community (Euratom).

The European Coal and Steel Community was established in 1952 by the Treaty of Paris 1951. The two Treaties of Rome 1957 established the European Economic Community and the European Atomic Energy Community. The

Treaty establishing the EEC was signed on 25 March 1957 by six member states – France, Germany, Italy, The Netherlands, Belgium and Luxemburg. In signing the document each of the member countries committed itself to the elimination of trade tariffs within the new Community and the development of a Common Market.

In 1963 the UK first applied for entry to the EEC and was vetoed by the French President, Charles de Gaulle. In 1967 the assemblies of the three communities were merged and a single European Commission and a single European Parliamentary Assembly were created. Britain reapplied for membership in 1967 and was again vetoed by de Gaulle. The resignation of de Gaulle in 1969 unlocked the door for discussions regarding British entry. On 1 January 1973 the UK, Denmark and Ireland became members of the EEC. The membership of the Community has since continued to enlarge and now includes Greece, Portugal, Spain, Austria, Finland and Sweden. The development of the European Union in 1993 brought with it a renaming of the EEC to the European Community (EC).

## The EC framework

The constitution of the Communities of the European Union is contained within Treaties which form part of the national law of member countries and impose supremacy of Community legislation over national law. The Treaties also serve to create and determine the authority of the institutions of the Communities in the development of Community legislation.

**The Treaty of Rome**   Article 2 of the Treaty of Rome 1957 states a principal aim of the European Community as being that of:

> . . . establishing a common market . . . to promote throughout the Community a harmonious development of economic activities, a continuous and balanced expansion, an increase in stability, an accelerated raising of the standards of living and closer relations between the states belonging to it.

In order to enable the effective functioning of such a common market Article 3(a) of the Treaty of Rome aims for:

> . . . the approximation of the law of member states to the extent required for the proper functioning of the Common Market.

In fulfilling this aim, Community states must ensure a parity of laws relevant to the functioning of the Common Market. The institutions of the European

Union provide for the development of necessary legislation to be imposed upon or adopted by Community states.

The legislation of the EU and the institutions party to the development of this legislation are now briefly outlined.

## Legislation of the European Union

There are three types of European legislation:

- Regulations
- Directives
- Decisions

Regulations are entirely binding upon all member states and form part of the law of member states. No further action is required by member states in undertaking to implement such law.

Directives specify that which has to be achieved by the national law of member states. They are binding and set goals and it is the responsibility of each member state to achieve these goals by a specified date. In this way Directives recognize the existence of differences within the legal systems of member states and facilitate the adoption of varying approaches to the achievement of specified goals.

Decisions are not directed at all member states, unlike Regulations and Directives. Decisions are made by the Council and are applied to individuals, corporations and specific member states.

A numbering system is employed to record and identify enacted legislation. A two-digit number identifies the year of publication with a further number identifying the chronology of publication within that year. The year is identified first for Decisions and Directives, while Regulations record the year second. Regulations are recorded thus '111/99'; Decisions and Directives are recorded thus '99/111'. The publication recording all Regulations and the majority of Directives is *Official Journal 'L' series*. The two sections of the journal present:

- Acts whose publication is compulsory – Regulations.
- Acts whose publication is not compulsory – Directives and Decisions.

## Institutions of the European Union

The institutions of the European Union work together to achieve the common objectives of the Treaties. Sovereignty to enact and enforce legislation is provided to these institutions by the Treaties. The composition of key institutions is now outlined. In presenting this outline the legislative processes of the Union are not described in detail as they are well documented.

### The Commission of the European Communities

The Commission proposes most Community legislation and consists of 20 commissioners each appointed by the governments of the member states. These appointments are subject to approval by the Parliament. Twenty-three Directorate-Generals support the commissioners with each Directorate-General being responsible for administration of different areas of policy.

The Commission ensures that Directives and Regulations are properly implemented by member states and can enforce these laws by bringing cases before the European Court of Justice.

### Council of Ministers of the European Union and the European Council

The Council of Ministers is constituted by 15 ministers, one minister representing each member state. The Council is responsible for enacting Union legislation. Each of the Council's 15 representatives is a member state government minister. Representation varies depending upon the particular nature of the issue or policy under discussion, for example Council discussions appertaining to transport would be attended by the relevant member state government minister responsible for that area. The European Council is a special meeting of the Council of Ministers and is constituted by the Head of Government/State of each member state.

### The European Parliament

The European Parliament has 626 elected representatives. The United Kingdom holds 87 of the 626 seats. Allegiances are not national ones but are determined by political affiliation across international boundaries. The legislative role of Parliament is limited to assisting in the drafting of Directives and Regulations and proposing amendments to the Commission for consideration.

### Economic and Social Committee

The Economic and Social Committee consists of 222 members, representing trade unions, employer organizations and other such social and economic activity nominated by governments. This Committee has no legislative power but is consulted before Decisions are taken on *inter alia* health and safety legislation.

### Court of Justice of the European Communities

This court is located in Luxemburg and is made up of 15 Judges. The Judges are appointed by agreement of the member states for a term of six years. Further to Article 164 of the Treaty of Rome the court interprets and ensures the application of Community law.

## Health and safety and its European development

**The Treaty of Rome:
health and safety**

With specific regard to health and safety, three of the Treaty of Rome's Articles provide for the European Community's regulation of health and safety within the Common Market. These are Articles 36, 100 and 235.

Article 36 of the Treaty allows for member states to restrict the free movement of goods among themselves. This restriction could be viewed as apparently contravening Article 30 of the Treaty. This Article prohibits 'quantitative restrictions on imports and all measures having equivalent effect ... between member states'. The Article 36 restriction on the trade of goods or products from member states or third parties is permitted if it is justified as being for, *inter alia* 'the protection of health and life of humans, animals and plants'.

Article 100 of the Treaty is concerned with promoting cooperation among member states and the functioning of the Common Market. Within the context of health and safety, more precisely the prevention of occupational accidents and diseases and occupational hygiene, Article 100 granted authority for the Council to issue opinions to, and arrange consultations with member states. Article 100 of the original Treaty of Rome clearly somewhat limited the Community's jurisdiction as regards health and safety.

Article 235 of the Treaty also provides the European Community with a basis for the jurisdiction of health and safety matters. Although never used, the Article allows for action 'necessary to attain, in the course of the operation of the Common Market, one of the objectives of the Community' (Treaty of Rome Article 235, 1973 version). Such action could potentially relate to health and safety. In order for such action to be undertaken to ensure one of the objectives of the Community the Commission would have to issue a proposal to the Council and this in turn would have to be unanimously accepted.

**The Single European
Act 1986 (1987
OJ (L169) 1.)**

The Single European Act, signed in February 1986, came into effect on 1 July 1987 and in doing so amended the Treaty of Rome. Key amendments to the Treaty of Rome concerned:

- the development of the internal market
- a change in the voting procedure of the Council
- the expression of Community influence in the area of social policy (health and safety).

These key amendments are briefly outlined below.

### The development of the internal market

Article 13 of the Act introduced a new Article 8A into the Treaty of Rome. A date was set by which the European Community had to complete the internal market structure. Article 8A stated that the internal market was to be established by 31 December 1992 and that:

> The internal market shall comprise an area without frontiers in which the free movement of goods, persons, services and capital is ensured in accordance with the provisions of this Treaty.

### A change in the voting procedure of the Council

The insertion of Article 100A into the Treaty of Rome brought about qualified majority voting for the Council. The definition of 'qualified majority' is provided in Article 148(2) of the Treaty of Rome and means 54 out of 76 Council votes. This equates to a little more than 70 per cent of the Council's total votes. Importantly, this means that no longer can one or two member states defeat a proposal relating to the internal market under qualified voting.

### The expression of Community influence in the area of social policy (health and safety)

The insertion of Article 118A into the Treaty of Rome provided for the expression of Community influence in the area of social policy. Within the context of the act 'social policy' includes health and safety. Article 118A directs member states to:

> pay particular attention to encouraging improvements, especially in the working environment, as regards the health and safety of workers, and set as their (the member states) objectives the harmonization of conditions in this area.

Article 118A also authorizes the Council to adopt by the means of Directives 'minimum requirements for gradual implementation' of improvements in the working environment as regard the health and safety of workers (Article 118(2)). In adopting Directives for Community health and safety-related minimum requirements the Council is able to act under a qualified majority on a Commission proposal, in cooperation with Parliament and after consultation of the Economic and Social Committee, to encourage improvements that will guarantee a better level of protection for the health and safety of workers, especially in the working environment.

**The Treaty on European Union and the Social Charter**

The 1991 Treaty on European Union, commonly referred to as the Maastricht Treaty, further amended the Treaty of Rome. The Treaty on European Union set out a detailed timetable for economic and monetary union and provided for the development of Community foreign and defence policies.

With regard to health and safety, specifically public health, the Treaty on European Union stated that the Community:

shall contribute towards ensuring a high level of human health protection by encouraging cooperation between member states and, if necessary, lending support to their action.

(Treaty of Rome as amended by the Treaty on European Union, Article 129)

In December 1987, prior to the Treaty on European Union, the Commission agreed a broad plan of social policy initiatives. This agreement was further to Article 118A of the Treaty of Rome and its expression of Community influence in the area of social policy. The rationale for such policy development was that it would serve the objective of Article 118A and in so doing would provide a social dimension to the development of a Single European Market. Social initiatives, it was considered, would among other things better facilitate perceptions of inclusion and participation within a Single European Market by people Community-wide. In this context a 'social charter' of fundamental rights was seen as a vehicle complementary to the acceptance of the Single European Market and its objective of business efficiency. The Charter, subject to unanimous acceptance, was to be ratified by the forthcoming Treaty on the European Union.

In December 1989 the Social Charter was adopted by all but one Head of State/Government at the Strasburg European Council (Conclusions of Strasburg European Council, EC Bulletin 12–1989, 1.1.1.). The UK did not adopt the Charter. A factor contributing to this decision was the Single European Act's introduction of the qualified majority voting procedure for proposed social policy-related legislation (Article 100A, Treaty of Rome). Such qualified majority voting now meant that any member country could find itself bound on matters of social policy (and health and safety) by an imposed agreement of the others.

The UK's decision not to adopt the Charter meant that, in turn, the Charter was not adopted into the Treaty on European Union but was instead attached as a protocol. As such the Social Charter possessed no legal status in itself. All member states, with the exception of the UK, agreed in an annex to the attached Protocol to instruct the European Commission to implement the Charter by the means of proposing Directives. While Community members were not bound to implement these Directives, agreement was that they would. The UK, although not party to the Charter, has nevertheless implemented social policy-related Directives (including those of health and safety) proposed and developed as a result of the Social Charter. Further to the signing of the Treaty of Amsterdam in October 1997 the UK is bound on matters of European Social Policy.

Article 2 of the Protocol on Social Policy is concerned with minimum requirements for gradual implementation of measures designed to improve health and safety at work and consultation of workers. In implementing the Charter the Commission submits legislative proposals to the Council of

Ministers. Further to adoption of proposed Directives member states are accorded responsibility for the guaranteeing of Charter rights. In undertaking to guarantee Charter rights and take forward Commission proposals member states implement necessary measures in accordance with their national practices.

One significant Commission proposal to be adopted was the commonly termed 'Framework Directive' (89/391/EEC) (EC, 1989) on the introduction of measures to encourage improvements in the safety and health of workers at work. This Directive was the first Directive proposed by the Commission, addressed to the member states, regarding the implementation of Article 118a of the EEC Treaty.

## The Framework Directive

On 12 June 1989 the Council of the European Communities adopted Directive 89/391/EEC – 'measures to encourage improvements in the safety and health of workers at work' – commonly referred to as the Framework Directive (EC, 1989). The Directive required compliance of member states by 31 December 1992 and addressed:

- the prevention of occupational risks
- the protection of safety and health
- the informing, consultation and training of workers and their representatives, and
- principles concerning the management of the above measures.

The Directive does not prescribe procedure but sets out duties of care and obligations that the employer must fulfil in the carrying out of work. These duties and obligations broadly concern management of the above aspects with a considered risk assessment-based approach.

Article 5 states the general duty of care of employers, this being that: 'the employer shall have a duty to ensure the safety and health of workers in every aspect related to work' (Directive 89/391/EEC [1989] OJ L183/1).

Obligations of the employer are extended to:

- workers from any outside undertaking and/or establishments engaged in work in his undertaking and/or establishment
- provide information and appropriate instruction.

The Directive also importantly provides for some common Community-wide definitions:

- An 'employer' is defined as 'any natural or legal person who has an employment relationship with the worker and has responsibility for the undertaking and/or establishment'.
- A 'worker' is defined as 'any person employed by an employer, including trainees and apprentices but excluding domestic servants'.

Appendix 1 (p. 227) details the Framework Directive and provides for the consultation of the 19 Articles.

Further European Directives have been adopted to develop the Framework Directive and support improvements in specific aspects of safety and health of workers at work. These Directives are commonly referred to as 'daughter Directives'. There have been 14 daughter Directives. A number of these are applicable to construction businesses, they include:

- The workplace (89/654/EEC)
- Use of work equipment (89/655/EEC)
- Use of personal protective equipment (89/656/EEC)
- Manual handling of loads (90/269/EEC [1990] OJ L56/9)
- Display screen equipment (90/270/EEC)
- Carcinogens (90/394/EEC [1990] OJ L196/1)
- Biological agents (90/679/EEC [1990] OJ L334/1)
- Temporary or mobile sites (92/57/EEC)
- Safety and health signs (92/58/EEC [1992] OJ L245/23)
- Pregnant workers (92/85/EEC [1992] OJ L348/1)
- Protection of young people at work (94/33/EEC).

While the workplace Directive excludes the location of the construction sites, it is applicable to offices and other such workplaces and hence off-site construction activity.

Within the UK the Framework Directive was implemented by the Management of Health and Safety at Work Regulations 1992 (MHSWR) (SI 1992 No. 2051) (HSE, 1992), which were revoked and replaced by the Management of Health and Safety at Work Regulations 1999 (HSE, 1999) (see Chapter 7). The associated daughter Directives have been implemented in the UK by the introduction of various relevant Regulations. The nature and implications of these Regulations and other UK health and safety law, implemented as a response to the requirements of adopted European Directives is presented in Chapters 7, 8 and 9. The reader is referred to these chapters for an analysis of health and safety law.

## Key points

This chapter has identified that:

- EC Treaties form a part of the national law of member countries.
- There are three types of European legislation: Regulations, Directives, and Decisions.
- The principal Directive influencing health and safety is Directive 89/391/EEC (EC, 1989) on the measures to encourage improvements in the safety and health of workers at work.

## Further reading

Chalmers, D. (1998) *European Union Law Vol 1: Law and EU Government*, Dartmouth Publishing, Hants.

Dalby, J. (1998) *EU Law for the Construction Industry*, Blackwell Science, Oxford.

Gale, S. (1998) *EC Law*, second edition, Butterworths, London.

Kapteyn, P. J. G. and VerLoren Van Thermaat, P. (1998) *Introduction to the Law of the European Community*, third edition, Kluwer Law International, London.

Lenaerts, K., Nuffel, P. and Bray, R. (1999) *Constitutional Law of the European Union*, Sweet & Maxwell, London.

Wetherill, S. and Beaumont, P. (1999) *EU Law*, third edition, Penguin, London.

## References

European Commission (EC) (1989) Directive 89/391/EEC on the introduction of measures to encourage improvements in the health and safety of workers at work, HMSO, London.

Health and Safety Executive (HSE) (1992) *The Management of Health and Safety at Work Regulations 1992*, HMSO, London.

Health and Safety Executive (HSE) (1999) *The Management of Health and Safety at Work Regulations 1999*, HMSO, London.

# 6 The Health and Safety at Work, etc. Act 1974

## Introduction

The Health and Safety at Work Act, etc. 1974 (HSE, 1974), commonly referred to as the HSWA, provides the basis for British health and safety law. This chapter presents an outline of key aspects of the Act and in so doing provides an insight into:

- The enabling framework of the Act that supports the implementation of European health and safety Directives.
- Key sections of the Act.
- The key duties imposed by the Act relating to the safeguarding of health and safety standards.
- Liability for breach of health and safety duty.
- The institutions of health and safety enforcement – the Health and Safety Commission and the Health and Safety Executive.
- The enforcement process and resulting penalties.

The 'further reading' section at the end of this chapter (p. 58) lists publications which focus on health and safety at work and legislation.

## The Act and its statutory provisions

The Health and Safety at Work, etc. Act came about as a response to constantly expanding, ever more detailed, UK health and safety law. The Act consolidated much legislation and provided for the development of a 'personal responsibility' approach to health and safety.

While providing primary health and safety legislation, the Act is an enabling Act and has a provision for the development of a framework of health and safety statutory instruments – 'Regulations' – and any associated standards and approved codes of practice. While standards and codes of practice are not legally binding – they are provided for by section 16 – compliance with such standards and codes is admissible to a court of law as evidence of satisfying the requirements of relevant Regulations.

It is section 15 of the Act that provides for the implementation of health and safety Regulations. As such section 15 – 'Health and Safety Regulations' – is a vehicle for the translation of European health and safety Directives into UK law. Examples of Regulations are presented in Chapters 7, 8 and 9.

## Key duties imposed by the Act and key sections relating to the safeguarding of health and safety standards

The Act imposes duties upon a number parties:

- Section 2 places general duties on employers towards employees.
- Section 3 places duties on employers and the self-employed to persons other than their employees.
- Section 4 places duties on people in control of premises.
- Section 6 places duties on people who design, manufacture, supply and install plant, equipment and substances used during a project.
- Section 7 places a duty on every employee.
- Section 8 places a duty on everybody.

These sections and their duties are described below.

**Key sections of the Act**

**Section 2: General duties of employers to their employees**

(1) It shall be the duty of every employer to ensure, so far as is reasonably practicable, the health, safety and welfare at work of all his employees.

(2) Without prejudice to the generality of an employer's duty under the preceding subsection, the matters to which that duty extends include in particular –

(a) the provision and maintenance of plant and systems of work that are, so far as is reasonable, safe and without risks to health

(b) arrangements for ensuring, so far as is reasonably practicable, safety and absence of risk to health in connection with the use, handling, storage and transport of articles and substances

(c) the provision of such information, instruction, training and supervision as is necessary to ensure, so far as is reasonably practicable, the health and safety at work of his employees

(d) so far as is reasonably practicable, as regards any place of work under the employer's control, the maintenance of it in a condition that is safe and without risks to health and safety and the provision and maintenance of means of access to and egress from it that are safe and without such risks

(e) the provision and maintenance of a working environment for his employees that is, so far as is reasonably practicable, safe, without risks to health, and adequate as regards facilities and arrangements for their welfare at work.

(3) Except in such cases as may be prescribed, it shall be the duty of every employer to prepare and as often as may be appropriate revise a written statement of his general policy with respect to the health and safety at work of his employees and the organization and arrangements for the time being in force for carrying out that policy, and to bring the statement and any revision of it to the notice of all of his employees.

(4) Regulations made by the Secretary of State may provide for the appointment in prescribed cases by recognized trade unions (within the meaning of the Regulations) of safety representatives from amongst the employees, and those representatives shall represent the employees in consultations with the employers under subsection (6) below and shall have such other functions as may be prescribed.

(5) . . .

(6) It shall be the duty of every employer to consult any such representatives with a view to the making and maintenance of arrangements which will enable him and his employees to cooperate effectively in promoting and developing measures to ensure the health and safety at work of the employees, and in checking the effectiveness of such measures.

(7) In such cases as may be prescribed it shall be the duty of every employer, if requested to do so by the safety representatives mentioned in [subsection (4)] above, to establish, in accordance with Regulations made by the Secretary of State, a safety committee having the function of keeping under review the measures taken to ensure the health and safety at work of his employees and such other functions as may be prescribed.

## Section 3: General duties of employers and self-employed to persons other than their employees

(1) It shall be the duty of every employer to conduct his undertaking in such a way as to ensure, so far as is reasonably practicable, that persons not in his employment who may be affected thereby are not thereby exposed to risks to their health and safety.

(2) It shall be the duty of every self-employed person to conduct his undertaking in such a way as to ensure, so far as is reasonably practicable that he and other persons (not being his employees) who may be affected thereby are not thereby exposed to risks to their health or safety.

(3) In such cases as may be prescribed, it shall be the duty of every employer and every self-employed person, in the prescribed circumstances and in the prescribed manner, to give to persons (not being his employees) who may be affected by the way in which he conducts his undertaking the prescribed information about such aspects of the way in which he conducts his undertaking as might affect their health or safety.

The case of *R. v Associated Octel Co. Ltd* [1996] (1 WLR 1543) provides for interpretation of section 3(1) with specific regard to independent contractors carrying out regular maintenance. The facts of the case were that an independent contractor was contracted to clean and maintain Octel's Chemical plant. This work was carried out during the plant's annual shutdown for maintenance. An employee of the contractor was badly burned in an explosion that occurred while cleaning and maintaining a tank. The contractor was convicted of an offence with regard to its employee under section 2 of the Act while Octel was also convicted under section 3(1). These convictions were appealed to the Court of Appeal and to the House of Lords and on both occasions the appeal was dismissed.

Importantly this case provides for an interpretation of section 3(1) with regard to the general duties of employers to maintenance contractors, (persons other than employees), who carry out maintenance and cleaning work necessary to the employer's conducting of business. In this instance the 'conducting of business' was the manufacture of chemicals. Further to the case, employers have a duty to stipulate conditions to avoid health and safety risks to independent maintenance contractors. Lord Hoffman, in rejecting the appeal to the House of Lords, stated that it is the obligation of the employer to take 'reasonably practicable steps to avoid risks to the contractors' servants which arise, not merely from the physical state of the premises . . . but also from the inadequacy of the arrangements which the employer makes with the contractor for how they do the work'.

### Section 4: General duties of persons concerned with premises to persons other than their employees

(1) This section has effect for imposing on persons duties in relation to those who –
   (a) are not their employees; but
   (b) use non-domestic premises made available to them as a place of work or as a place where they may use plant or substances provided for their use there

and applies to premises so made available and other non-domestic premises used in connection with them.

(2) It shall be the duty of each person who has, to any extent, control of premises to which this section applies or of the means of access thereto or egress therefrom or of any plant or substance in such premises to take such measures as it is reasonable for a person in his position to take to ensure, so far as is reasonably practicable, that the premises, all means of access thereto or egress therefrom available for use by persons using the premises, and any plant or substance in the premises or, as the case may be, provided for use there, is or are safe and without risks to health.

(3) Where a person has, by virtue of any contract or tenancy, an obligation of any extent in relation to –

(a) the maintenance or repair of any premises to which this section applies or any means of access thereto or egress therefrom; or

(b) the safety of or the absence of risks to health arising from plant or substances in any such premises

that person shall be treated, for the purposes of subsection (2) above, as being a person who has control of the matters to which his obligation extends.

(4) Any reference in this section to a person having control of any premises or matter is a reference to a person having control of the premises or matter in connection with the carrying on by him of a trade, business or other undertaking (whether for profit or not).

## Section 6: General duties of designers, manufacturers and suppliers of articles and substances for use at work

(1) It shall be the duty of any person who designs, manufactures, imports or supplies any article for use at work or any article of fairground equipment –

(a) to ensure, so far as is reasonably practicable, that the article is so designed and constructed that it will be safe and without risks to health at all times when it is being set, used, cleaned or maintained by a person at work

(b) to carry out or arrange for the carrying out of such testing and examination as may be necessary for the performance of the duty imposed on him by the preceding paragraph

(c) to take such steps as are necessary to secure that persons supplied by that person with the article are provided with adequate information about the use for which the article is designed or has been tested and about any conditions necessary to ensure that it will be safe and without risks to health at all times as are mentioned in paragraph (a) above and when it is being dismantled or disposed of; and

(d) to take such steps as are necessary to secure, so far as is reasonably practicable, that persons so supplied are provided with all such revisions of information provided to them by virtue of the preceding paragraph as are necessary by reason of it becoming known that anything gives rise to a serious risk to health and safety.

(1A) It shall be the duty of any person who designs, manufactures, imports or supplies any article of fairground equipment –

(a) to ensure, so far as is reasonably practicable, that the article is so designed and constructed that it will be safe and without risks to health at all times when it is being used for or in connection with the entertainment of members of the public

(b) to carry out or arrange for the carrying out of such testing and examination as may be necessary for the performance of the duty imposed on him by the preceding paragraph

(c) to take such steps as are necessary to secure that persons supplied by that person with the article are provided with adequate information about the use for which the article is designed or has been tested and about any conditions necessary to ensure that it will be safe and without risks to health at all times when it is being used for in connection with the entertainment of members of the public; and

(d) to take such steps as are necessary to secure, so far as is reasonably practicable, that persons so supplied are provided with all such revisions of information provided to them by virtue of the preceding paragraph as are necessary by reason of its becoming known that anything gives rise to a serious risk to health or safety.

(2) It shall be the duty of any person who undertakes the design or manufacture of any article for use at work or of any article of fairground equipment to carry out or arrange for the carrying out of any necessary research with a view to the discovery and, so far as is reasonably practicable, the elimination or minimization of any risks to health or safety to which the design or article may give rise.

(3) It shall be the duty of any person who erects or installs any article for use at work in any premises where that article is to be used by persons at work or who erects or installs any article of fairground equipment to ensure, so far as is reasonably practicable, that nothing about the way in which the article is erected or installed makes it unsafe or a risk to health at any such time as is mentioned in paragraph (a) of subsection (1) or, as the case may be, in paragraph (a) of subsection (1) or (1A) above.

(4) It shall be the duty of any person who manufactures, imports or supplies any substance –

    (a)  to ensure, so far as is reasonably practicable, that the substance will be safe and without risks to health at all times when it is being used, handled, processed, stored or transported by a person at work or in premises to which section (4) above applies

    (b)  to carry out or arrange for the carrying out of such testing and examination as may be necessary for the performance of the duty imposed on him by the preceding paragraph

    (c)  to take such steps as are necessary to secure that persons supplied by that person with the substance are provided with adequate information about any risks to health or safety to which the inherent properties of the substance may give rise, about the results of any relevant tests which have been carried out on or in connection with the substance and about any conditions necessary to ensure that the substance will be safe and without risks to health at all such times as are mentioned in paragraph (a) above and when the substance is being disposed of; and

    (d)  to take such steps as are necessary to secure, so far as is reasonably practicable, that persons so supplied are provided with all such revisions of information provided to them by virtue of the preceding paragraph as are necessary by reason of its becoming known that anything gives rise to a serious risk to health and safety.

(5)  It shall be the duty of any person who undertakes the manufacture of any substance to carry out or arrange for the carrying out of any necessary research with a view to the discovery and, so far as is reasonably practicable, the elimination or minimization of any risks to health or safety to which the substance may give rise at all such times as are mentioned in paragraph (a) of subsection (4) above.

(6)  Nothing in the preceding provisions of this section shall be taken to require a person to repeat any testing, examination or research which has been carried out otherwise than by him or at his instance, in so far as is it is reasonable for him to rely on the results thereof for the purposes of those provisions.

(7)  Any duty imposed upon any person by any of the preceding provisions of this section shall extend only to things done in the course of a trade, business or other undertaking carried on by him (whether for profit or not) and to matters within his control.

(8)  Where a person designs, manufactures, imports or supplies an article for use at work or an article of fairground equipment and does so for or to another on the basis of a written undertaking by that other to take specified steps sufficient to ensure, so far as is reasonably practicable, that the article will be safe and without risks to health at all such times as are mentioned in paragraph (a) of subsection (1) or, as the case may be, in paragraph (a) of subsection (1) or (1A) above, the

undertaking shall have the effect of relieving the first-mentioned person from the duty imposed by virtue of that paragraph above to such an extent as is reasonable having regard to the terms of the undertaking.

(8A) Nothing in subsection (7) or (8) above shall relieve any person who imports any article or substance from any duty in respect of anything which –

(a) in the case of an article designed outside the United Kingdom, was done by and in the course of any trade profession or other undertaking carried on by, or was within the control of, the person who designed the article; or

(b) in the case of an article or substance manufactured outside the United Kingdom, was done by and in the course of any trade, profession or other undertaking carried on by, or was within the control of, the person who manufactured the article or substance.

(9) Where a person ('the ostensible supplier') supplies any article or substance to another ('the customer') under a hire-purchase agreement, conditional sale agreement or credit-sale agreement, and the ostensible supplier –

(a) carries on the business of financing the acquisition of goods by others by means of such agreement; and

(b) in the course of that business acquired his interest in the article or substance supplied to the customer as a means of financing its acquisition by the customer form a third person ('the effective supplier')

the effective supplier and not the ostensible supplier shall be treated for the purposes of this section as supplying the article or substance to the customer, and any duty imposed by the preceding provisions of this section on suppliers shall accordingly fall on the effective supplier and not on the ostensible supplier.

(10) For the purposes of this section an absence of safety or a risk to health shall be disregarded in so far as the case in or in relation to which it would arise is shown to be one the occurrence of which could not reasonably be foreseen; and in determining whether any duty imposed by virtue of paragraph (a) of subsection (1), (1A) or (4) above has been performed regard shall be had to any relevant information or advice which has been provided to any person by the person by whom the article has been designed, manufactured, imported or supplied or, as the case may be, by the person by whom the substance has been manufactured, imported or supplied.

### Section 7: General duties of employees at work

It shall be the duty of every employee while at work –

(a)  to take reasonable care for the health and safety of himself and of other persons who may be affected by his acts or omissions at work; and

(b)  as regards any duty or requirement imposed on his employer or any other by or under any of the relevant statutory provisions, to cooperate with him so far as is necessary to enable that duty or requirement to be performed or complied with.

### Section 8: Duty not to interfere with or misuse things provided pursuant to certain provisions

No person shall intentionally or recklessly interfere with or misuse anything provided in the interests of health, safety or welfare in pursuance of any of the relevant statutory provisions.

**The key duties of health and safety law**

The key duties imposed by health and safety law fall within three categories:

1.  *Absolute* – this is a duty that *must* be carried out. It imposes an absolute obligation on a party and any breach of duty may result in prosecution.
2.  *Practicable* – this is a duty that should be carried out irrespective of inconvenience, time or cost. The standard of performance is a high standard, but not absolute.
3.  *Reasonably practicable* – this is a duty that is carried out having considered the balance of that duty against inconvenience and cost involved. Where cases of breach of duty are brought, it is the responsibility of the accused to demonstrate that it was not reasonably practical to have done more than that done to comply with the duty.

Duties that must be carried out 'so far as reasonably practicable' occur frequently throughout the HSWA. The Health and Safety Executive provide further interpretation of the term 'so far as is reasonably practicable':

Someone who is required to do something so far as is reasonably practicable, must assess, on the one hand, the risks of a particular work activity or environment, and, on the other hand, the physical difficulties, time, trouble and expense which would be involved in taking steps to avoid the risks. If, for example, the risks to health and safety of a particular work process are very low, and the cost or technical difficulties of taking certain steps to avoid those risks are very high, it might not be reasonably practicable to take those steps. However, if the risks are very high, then less weight can be given to the cost of measures needed to avoid those risks. The comparison does not include the financial standing of the employer. A precaution which is 'reasonably practicable' for a prosperous employer is equally 'reasonably practicable' for the less well off. The phrase so far as is reasonably practicable, without the word reasonably, implies a strict standard.

(HSE, para 23)

## Liability for breach of health and safety duty

The actions arising from a breach of a duty imposed by the HSWA and its associated Regulations are as follows:

*   A breach of any duty imposed by sections 2, 3, 4 and 7 of the HSWA is a criminal offence. Section 47 of the Act forbids civil actions for statutory breach of the Act in instances of prosecution under the act, when

    ... failing to comply with a duty so far as is reasonably practicable, it is up to the accused to show the court that it was not practicable or not reasonably practicable (as appropriate) for him to do more than he had in fact done to comply with the duty.

    (HSE, para 24)

*   Non-compliance with health and safety Regulations (issued under section 15 of the Act) is a criminal offence. Civil action may also be brought, unless expressly excluded, for breach of a duty imposed by health and safety Regulations.

Section 17 of the Act provides for the use of approved codes of practices in criminal proceedings.

### Section 17: Use of approved codes of practice in criminal proceedings

(1)   A failure on the part of any person to observe any provision of an approved code of practice shall not of itself render him liable to any civil or criminal proceedings; but where in any criminal proceedings a party is alleged to have committed an offence by reason of a contravention of any requirement or prohibition imposed by or under any such provision as is mentioned in section 16(1) being a provision for which there was an approved code of practice at the time of the alleged contravention, the following subsection shall have effect with respect to that code in relation to those proceedings.

(2) Any provision of the code of practice which appears to the court to be relevant to the requirement or prohibition alleged to have been contravened shall be admissible in evidence in the proceedings; and if it is proved that there was at any material time a failure to observe any provision of the code which appears to the court to be relevant to any matter which it is necessary for the prosecution to prove in order to establish a contravention of that requirement or prohibition, that matter shall be taken as proved unless the court is satisfied that the requirement or prohibition was in respect of that matter complied with otherwise than by way of observance of that provision of the code.

(3) In any criminal proceedings –

(a) a document purporting to be a notice issued by the Commission under section 16 shall be taken to be such a notice unless the contrary is proved; and

(b) a code of practice which appears to the court to be the subject of such a notice shall be taken to be the subject of that notice unless the contrary is proved.

## Enforcement of the Act: the institutions

Sections 10–14 of the Act have created and conferred powers upon the Health and Safety Commission (HSC) and the Health and Safety Executive (HSE).

The Commission is a corporate body which, further to Section 11(2) of the Act, serves to:

- Assist and encourage matters relevant to the promote of health and safety.
- Undertake and publish research.
- Provide an information and advisory service.
- Submit proposals, from time to time, regarding the making of relevant Regulations.
- Give effect to any Directions given to it by the Secretary of State.

The composition of the Commission is determined by Section 10(3) of the Act and as such consists of up to 10 people:

- A chairperson – appointed by the Secretary of State for Employment.
- 6–9 members – 3 each from employer and employee organizations and 3 other people chosen by the Secretary of State from local authorities and professional bodies.

Expert advice regarding health and safety in the construction industry is provided to the Commission by an advisory committee – the Construction Industry Advisory Committee – and by the Health and Safety Executive.

The Executive, like the Commission, is a corporate body and functions on behalf of the Commission. Further to Section 11(4) and (5), the Executive serves:

- to undertake the Commission's functions as directed by the Commission; and
- to give effect to such directions given by the Commission.

In effect the Executive is the operational arm of the Commission.

The composition of the Executive Commission is determined by Section 10(5) of the Act and as such consists of:

- Three persons – all appointed by the Commission, one with the approval of the Secretary of State.

## The enforcement process

Enforcement of the Act is undertaken by HSE inspectors and by local authority inspectors. All inspectors have available to them the same mechanisms of enforcement. It is the nature of the main activity of the business that determines the enforcing authority. The Health and Safety (Enforcing Authority) Regulations 1998, detail those premises and activities subject to inspection by local authority and HSE inspectors. Schedule 1 of the Regulations outlines those premises and activities which are a local authority's responsibility and schedule 2 outlines those premises and activities which are the HSE's responsibility – *inter alia*, building sites, factories and manufacturing, nuclear installations, railways, schools and hospitals. For further detail the reader is referred to these Regulations.

When considering an action, the inspector uses discretion and considers:

- the risk presented
- the gravity of the alleged offence
- the history of the business regarding previous compliance
- the inspector's confidence in the management of the business
- the likely effectiveness of a particular action.

(Croner, 1997)

In undertaking to ensure the enforcement of the Act and the maintenance of standards an inspector can:

- Inform and provide advice.
- Issue a letter of intention to serve a notice.
- Serve an Improvement Notice.
- Serve a Prohibition Notice.
- Prosecute.

The decision to prosecute rests with the enforcing authority and, should a breach of duty be established, can result in penalties, outlined by the Health and Safety Commission (HSC, 1995) as follows:

In the lower courts:

- For failure to comply with an improvement or Prohibition Notice, or court remedy order: a fine of up to £20,000, or six months' imprisonment, or both.
- For breaches of Sections 2–6 of the Health and Safety at Work, etc. Act 1974: a fine of up to £20,000.
- For other breaches of the Act not specified above, or any relevant statutory provisions under the Act: a fine of up to £5,000.

In higher courts:

- For failure to comply with an Improvement or Prohibition Notice, or court remedy order: 2 years' imprisonment, or an unlimited fine, or both.
- For contravening licence requirements or provisions relating to explosives: 2 years' imprisonment, or an unlimited fine, or both.
- For breaches of the Health and Safety at Work, etc. Act 1974, or of relevant statutory provisions under the Act: unlimited fines.

It is worth highlighting from the above identified penalties that the breach of Regulations made under the Act – a statutory provision – will result in a fine of up to £5,000 in a lower court and unlimited fines in a higher court. Details of such UK Health and Safety Regulations implemented via section 15 of the Act are presented within Chapters 7, 8 and 9.

## Key points

This chapter has identified that:

- The Health and Safety at Work, etc. Act 1974 (HSE, 1974) is the principal Act enabling the development of a framework for health and safety statutory instruments, or Regulations.

- The Act imposes duties on employers, the self-employed, employees and other persons in the workplace.
- Any breach of the Act is a matter which may have civil or criminal liability in law.

## Further reading

Dewis, M. (1998) *Tolley's Health and Safety at Work Handbook,* tenth edition, Tolley Publishing Company Ltd., Croydon.

Frederick Place Chambers (1995) *Health and Safety at Work: Legislation and Cases,* CLT Professional Cases, Birmingham.

Health and Safety Executive (HSE) (1984) *Working with Employers HSE35.*

Smith, P. (ed.) (1981) *Croner's Health and Safety at Work,* Croner Publications Ltd., Kingston upon Thames, Surrey.

Tullet, S. (ed.) (1979) *Croner's Health and Safety at Work,* Croner Publications Ltd., Kingston upon Thames, Surrey.

## References

Croner (1997) *The Role of Local Authorities in the Enforcement of Health and Safety, Croner's Health and Safety Briefing,* Issue No. 129, p. 5, 14 July 1997, Croner Publications Ltd., Kingston upon Thames, Surrey.

Health and Safety Commission (HSC) (1995) *Enforcement Policy Statement, Health and Safety Commission, Leaflet No. MISC030,* October 1995.

Health and Safety Executive (HSE) (1974) *The Health and Safety at Work, etc. Act 1974,* HMSO, London.

Health and Safety Executive (HSE) (1975) *A Guide to the Health and Safety at Work Act – Health and Safety series booklet HS(R)6,* HMSO, London.

# 7 The Management of Health and Safety at Work Regulations 1999

## Introduction

This chapter briefly outlines aspects of the Management of Health and Safety at Work Regulations 1999 (HSE, 1999). It provides an insight into:

- The basis for, and developments brought about by, the Regulations.
- The structure of the Regulations.
- The duties of employers and employees under the Regulations.
- 'Miscellaneous', non-duty, Regulations contained within the instrument.
- The undertaking of risk assessment.

## The basis and development of the Regulations

The Management of Health and Safety at Work Regulations (MHSWR) first came into effect on 1 January 1993 and were developed to implement the general provisions of the European Framework Directive (89/391/EEC) – 'measures to encourage improvements in the safety and health of workers at work' (EC, 1989). As such the Management of Health and Safety at Work Regulations 1992 (HSE, 1992) developed UK health and safety management with the bringing about of the provisions of risk assessment and management, health surveillance, and the appointment of competent health and safety assistants.

The Management of Health and Safety at Work Regulations 1992 were later amended by:

- The Management of Health and Safety at Work (Amendment) Regulations 1994 (SI 1994 No. 286).
- The Health and Safety (Young Persons) Regulations 1997 (SI 1997 No. 135).
- The Fire Precautions (Workplace) Regulations 1997 (SI 1997 No. 1840).

The development of each of these three amending statutory instruments was to implement accordingly, the European Pregnant Workers Directive, the Young Workers and the Framework Directive provisions regarding fire precautions.

## Revocation of the 1992 Regulations and developments brought about by the 1999 Regulations

On 29 December 1999 the 1992 Regulations were revoked and replaced by the Management of Health and Safety at Work Regulations 1999 (HSE, 1999). As a result of the introduction of the MHSW Regulations 1999 the following Regulations were also revoked:

- The Management of Health and Safety at Work (Amendment) Regulations 1994 (SI 1994 No. 286).
- The Health and Safety (Young Persons) Regulations 1997 (SI 1997 No. 135).
- The Fire Precautions (Workplace) Regulations 1997 (SI 1997 No. 1840).

In addition to these revocations, 16 statutory instruments have been amended in minor ways. These amended instruments are outlined later in this chapter in Regulation 29.

The MHSW Regulations 1999 have introduced the following into the Regulations:

- Principles of prevention that are to be applied where an employer implements any preventive and protective measures (Regulation 4).
- Specification that where possible employers use competent employees in preference to external sources for competent health and safety advice and assistance (Regulation 7(8)).
- The requirement for employers to arrange and ensure necessary contacts with external services, particularly as regards first aid, emergency medical care and fire fighting (Regulation 9).
- The requirement for the assessment of risks relating to:
  – new and expectant mothers at work (Regulation 16)
  – young persons at work (Regulation 19).
- An explicit statement that employers have no defence for contravention of their duties as a result of the acts or omissions of their employees or appointed competent persons (Regulation 21).

## Structure

The Management of Health and Safety at Work Regulations 1999 consist of 30 Regulations. A Health and Safety Commission approved code of practice and guidance (HSE, 2000) supports the Regulations. This approved code of practice and guidance provides practical guidance to the Regulations. It is not the intention of this chapter to detail such guidance, instead an overview of key aspects of the Regulations is presented.

The Regulations are:

| | |
|---|---|
| Regulation 1 | Citation, commencement and interpretation |
| Regulation 2 | Disapplication of the Regulations |
| Regulation 3 | Risk assessment |
| Regulation 4 | Principles of prevention to be applied |
| Regulation 5 | Health and safety arrangements |
| Regulation 6 | Health surveillance |
| Regulation 7 | Health and safety assistance |
| Regulation 8 | Procedures for serious and imminent danger and danger areas |
| Regulation 9 | Contacts with external services |
| Regulation 10 | Information for employees |
| Regulation 11 | Cooperation and coordination |
| Regulation 12 | Persons working in host employers' or self-employed persons' undertakings |
| Regulation 13 | Capabilities and training |
| Regulation 14 | Employees' duties |
| Regulation 15 | Temporary workers |
| Regulation 16 | Risk assessment in respect of new or expectant mothers |
| Regulation 17 | Certificate from a registered medical practitioner in respect of new or expectant mothers |
| Regulation 18 | Notification by new or expectant mothers |
| Regulation 19 | Protection of young persons |
| Regulation 20 | Exemption certificates |
| Regulation 21 | Provision as to liability |
| Regulation 22 | Exclusion of civil liability |
| Regulation 23 | Extension outside Great Britain |
| Regulation 24 | Amendment of the Health and Safety (First Aid) Regulations 1981 |
| Regulation 25 | Amendment of the Offshore Installations and Pipeline Works (First Aid) Regulations 1989 |
| Regulation 26 | Amendment of the Mines Miscellaneous Health and Safety Provisions Regulations 1995 |
| Regulation 27 | Amendment of the Construction (Health, Safety and Welfare) Regulations 1996 |
| Regulation 28 | Regulations to have effect as health and safety Regulations |
| Regulation 29 | Revocations and consequential amendments |
| Regulation 30 | Transitional provision |

## Employers' and employees' duties

Employers' duties are outlined in Regulations 3–19, with the exceptions of Regulations 14 and 18 which respectively outline the duties of employees and new and expectant mothers. These Regulations are outlined below.

### Regulation 3 – Risk assessment

A duty is placed on all employers and self-employed persons by this Regulation. This duty is to assess the risks to workers and any other persons who may be affected by undertakings. The significant findings of such an assessment must be recorded where employers or self-employed persons employ five or more employees. The Regulation also outlines the risk assessment requirements of employing a young person.

### Regulation 4 – Principles of prevention to be applied

This Regulation requires employers and the self-employed to introduce measures to control risks identified by a risk assessment. The Regulation states that:

> Where an employer implements any preventive and protective measures he shall do so on the basis of the principles in Schedule 1 of the Regulations.

Schedule 1 specifies the general principles of prevention set out in Article 6(2) of the Council Directive 89/391/EEC (EC, 1989), these being:

(a)  avoiding risks

(b)  evaluating the risks which cannot be avoided

(c)  combating the risks at source

(d)  adapting the work to the individual, especially as regards the design of workplaces, the choice of work equipment and the choice of working and production methods, with a view, in particular, to alleviating monotonous work and work at a predetermined work-rate and to reducing their effect on health

(e)  adapting to technical progress

(f)  replacing the dangerous by the non-dangerous or the less dangerous

(g)  developing a coherent overall prevention policy which covers technology, organization of work, working conditions, social relationships and the influence of factors relating to the working environments

(h)  giving collective protective measures priority over individual protective measures; and

(i)  giving appropriate instructions to employees.

### Regulation 5 – Health and safety arrangements

This Regulation requires employers to have health and safety arrangements in place:

4(1) Every employer shall make and give effect to such arrangements as are appropriate, having regard to the nature of his activities and the size of his undertaking, for the effective planning, organization, control, monitoring and review of the preventive and protective measures.

Such arrangements are to be recorded where employers or self-employed persons employ five or more employees.

### Regulation 6 – Health surveillance

This Regulation requires all employers to provide health surveillance to those employees identified as having risks to their health and safety as a result of their work.

The Regulations do not provide a definition of 'health surveillance', though the approved code of practice and guidance (HSE, 2000) outlines the minimum requirements for health surveillance as 'the keeping of a health record'.

The approved code of practice and guidance also states that health surveillance is to be undertaken when:

(a) there is an identifiable disease or adverse health condition related to the work concerned; and

(b) valid techniques are available to detect indications of the disease or condition; and

(c) there is a reasonable likelihood that the disease or condition may occur under the particular conditions of work; and

(d) surveillance is likely to further the protection of the health and safety of the employees to be covered.

Whatever the nature of health surveillance it is clear that such surveillance has to be appropriate to the risks identified by the risk assessment. In undertaking health surveillance it should be remembered that:

The primary benefit, and therefore the objective of health surveillance should be to detect adverse health effects at an early stage, thereby enabling further harm to be prevented. The result of health surveillance can provide a means of:

(a) checking the effectiveness of control measures

(b) providing feedback on the accuracy of the risk assessment; and

(c) identifying and protecting individuals at increased risk because of the nature of their work.

(HSE, 2000)

### Regulation 7 – Health and safety assistance

This Regulation requires that all employers appoint one or more competent persons to assist in undertaking to comply with the requirements and prohibitions of relevant statutory provisions and by Part II of the Fire Precautions (Workplace) Regulations 197. Paragraph (8) of the Regulation states that:

> Where there is a competent person in the employer's employment, that person shall be appointed . . . in preference to a competent person not in his employment.

The requirement for health and safety assistance does not apply to the self-employed employer who is not in partnership with any other person and where that person has 'sufficient training and experience or knowledge and other qualities properly to undertake the measures . . .'.

Paragraph (7) of this Regulation states that the requirement for health and safety assistance does not apply to individuals who are employers and are carrying on business in partnership

> where at least one of the individuals concerned has sufficient training and experience or knowledge and other qualities –
>
> (a) properly to undertake the measures he needs to take to comply with the requirements and prohibitions imposed upon him by or under the relevant statutory provisions; and
>
> (b) properly to assist his fellow partners in undertaking the measures they need to take to comply with the requirements and prohibitions imposed upon them by or under the relevant statutory provisions.

### Regulation 8 – Procedures for serious and imminent danger and for danger areas

This Regulation requires that employers establish procedures for employees to follow in the event of serious or imminent danger and that a sufficient number of competent persons be nominated to implement evacuation procedures.

For this purpose 'competent persons' are defined in paragraph (2)(c) of this Regulation to be those persons with 'sufficient training and experience or knowledge and other qualities to enable him properly to implement the evacuation procedures'.

### Regulation 9 – Contacts with external services

This Regulation states that 'every employer shall ensure that any necessary contacts with external services are arranged, particularly as regards first aid, emergency medical care and rescue work'.

With regard to contacts with external services the approved code of practice provides guidance and states, 'This may only mean making sure that that employees know the necessary telephone numbers and, where there is a significant risk, that they are able to contact any help they need'.

### Regulation 10 – Information for employees

This Regulation requires that employers provide employees with information on health and safety risks identified by assessment; preventive and protective health and safety measures; the identity of competent persons nominated to implement the procedures in the event of serious or imminent danger; and the risks arising from the conduct and undertaking of other employers and their employees in the instance of shared workplaces such as construction sites – this is identified as a requirement by Regulation 11(1)(c).

Regulation 10 also requires every employer, before employing a child, to provide a parent of the child with comprehensible and relevant information on health and safety risks identified by assessment, preventive and protective health and safety measures, and the risks arising from the conduct and undertaking of other employers and their employees in the instance of shared workplaces – this is identified as a requirement by Regulation 11(1)(c).

Further to the Health and Safety Information for Employers Regulations 1989 (HSE, 1989), amended by the Health and Safety Information for Employees (Modifications and Repeals) Regulations 1995 (HSE, 1995), employers have a legal duty to display the health and safety law poster in a prominent place or alternatively to provide employees with an equivalent leaflet. Since 1 October 1999 the poster has included an outline of the key requirements of the Management of Health and Safety at Work Regulations.

### Regulation 11 – Cooperation and coordination

This Regulation concerns employers who share a workplace with other employers or self-employed persons and requires that employers cooperate with each other in respect of the implementation of relevant statutory health and safety provisions and Part II of the Fire Precautions (Workplace) Regulations 1997, and, take all reasonable steps to coordinate with each other in respect of measures taken to comply with relevant statutory health and safety legislation and Part II of the Fire Precautions (Workplace) Regulations 1997.

Paragraph 1(c) requires that employers who share workplaces 'take all reasonable steps to inform the other employers concerned of risks to their employee's health and safety arising out of or in connection with the conduct by him of his undertaking'.

### Regulation 12 – Persons working in host employers' or self-employed persons' undertakings

This Regulation requires that all employers and self-employed persons provide to other employers and their employees working on site information regarding risks and measures taken in respect of the health and safety of such other employers and their employees.

Paragraph (4) of the Regulation states that every employer shall

(a)   ensure that the employer of any employees from an outside undertaking who are working in his undertaking is provided with sufficient information to enable that second-mentioned employer to identify any person nominated by that first-mentioned employer in accordance with Regulation 8(1)(b) to implement evacuation procedures as far as those employees are concerned; and

(b)   take all reasonable steps to ensure that any employees from an outside undertaking who are working in his undertaking receive sufficient information to enable them to identify any person nominated by him in accordance with Regulation 8(1)(b) to implement evacuation procedures as far as they are concerned.

This applies to self-employed persons working in an employer's undertaking in the same manner as it applies to employees from an outside undertaking.

Regulation 8(1)(b) concerns the nomination of a sufficient number of competent persons to implement evacuation procedures in the event of serious and imminent danger.

### Regulation 13 – Capabilities and Training

This Regulation requires every employer to consider the health and safety capabilities of employees when entrusting tasks. Employees are to be provided with adequate health and safety training upon their recruitment and on occasions where they are exposed to new or increased risks because of –

(2)(b)

(i)    their being transferred or given changes of responsibility within the employer's undertaking

(ii)   the introduction of new work equipment into or a change respecting work equipment already in use within the employer's undertaking

(iii)  the introduction of new technology into the employer's undertaking, or

(iv)   the introduction of new systems of work into or a change respecting a system of work already in use within the employer's undertaking.

Such training is to take place during working hours and is, where appropriate, to be repeated periodically and adapted to take account of new or changed risks.

### Regulation 14 – Employees' Duties

Under section 7 of the Health and Safety at Work, etc. Act 1974 (HSE, 1974) every employee has a duty to take reasonable care for the health and safety of themselves and others who may be affected by their acts or omissions at work. This Regulation places a duty on employees to utilize work items and undertake production processes correctly and in accordance with any safety training and instruction received.

This Regulation also states that it is the duty of every employee to inform his employer and employer's health and safety representative of:

- Any work situation which a trained and instructed employee reasonably considers to be a serious and immediate health and safety risk.
- Any matter which a trained and instructed employee reasonably considers to represent a shortcoming in the employer's protection arrangements for health and safety.

### Regulation 15 – Temporary workers

This Regulation requires that employers provide people on fixed-term contracts with:

- comprehensive information on requisite occupational skills and qualifications to carry out work safely; and
- health surveillance information relevant to the position.

### Regulations 16–18

These Regulations relate specifically to the management of health and safety of new and expectant mothers. Regulation 16 relates to employers' 'Risk assessment in respect of new or expectant mothers'. Regulation 17 relates to the suspension from work of a new or expectant mother who works at night 'for so long as is necessary for her health or safety'. This is done on the production of a Certificate from a registered medical practitioner or registered midwife. Regulation 18 states that action by the employer under Regulation 16 is not required until a new or expectant mother has 'notified the employer in writing that she is pregnant, has given birth within the previous six months, or is breastfeeding'.

### Regulation 19 – Protection of young persons

This Regulation relates to the protection of young persons from 'any risks to their health and safety which are a consequence of their lack of experience, or absence of awareness of existing or potential risks or the fact that young persons have not yet fully matured'.

## Miscellaneous regulations

Further to employer and employee duties required by the MHSW Regulations 1999 a number of 'miscellaneous' Regulations are outlined below.

### Regulation 20 – Exemption certificates

This provides for the Secretary of State for Defence to exempt by a certificate in writing any home or visiting forces or headquarters from the requirements of the Regulations with the exception of Regulations 16–18. The Regulation also allows the Secretary of State for Defence to exempt any member of the home or visiting forces or headquarters from the requirements imposed by Regulation 14.

### Regulation 21 – Provisions as to liability

Nothing in the relevant statutory provisions shall operate so as to afford an employer a defence in any criminal proceedings for a contravention of those by reason of any act or default of –

(a)  an employee of his, or

(b)  a person appointed by him under Regulation 7.

### Regulation 22 – Exclusion of civil liability

(1)  Breach of a duty imposed by these Regulations shall not confer a right of action in any civil proceedings.

(2)  Paragraph (1) shall not apply to any duty imposed by these Regulations on an employer –

(a)  to the extent that it relates to risk referred to in Regulation 16(1) to an employee; or

(b)  which is contained in Regulation 19.

### Regulation 23 – Extension outside Great Britain

The Regulations apply to offshore installations, wells, pipelines and pipeline works, and associated activities within Great Britain's territorial waters. Paragraph (2) of the Regulation ensures that offshore workers are protected while off duty.

### Regulation 24 – Amendment of the Health and Safety (First Aid) Regulations 1981

This Regulation revokes Regulation 6 of the Health and Safety (First Aid) Regulations 1981 and thereby revokes the HSE's powers of granting exemptions from the Health and Safety (First Aid) Regulations 1981.

### Regulations 25 and 26

These respectively amend the Offshore Installations and Pipeline Works (First Aid) Regulations 1989, and the Mines Miscellaneous Health and Safety Provisions Regulations 1995. The reader is directed to the Regulations for detail of these amendments.

## Regulation 27 – Amendment of the Construction (Health, Safety and Welfare) Regulations 1996

This Regulation provides for the amendment of paragraph (2) of Regulation 20 of the CHSW Regulations 1998 – arrangements for dealing with foreseeable emergencies on construction sites.

## Regulation 28 – Regulations to have effect as health and safety Regulations

This Regulation gives effect to the MHSW Regulations 1999 as provided for by Part 1 of the Health and Safety at Work, etc. Act 1974 – Health Safety and Welfare in Connection with Work, and Control of Dangerous Substances and Certain Emissions into the Atmosphere. This is subject to Regulation 9 of the Fire Precautions (Workplace) Regulations 1997.

## Regulation 29 – Revocations and consequential amendments

This Regulation serves to revoke the following Regulations:

- The Management of Health and Safety at Work Regulations 1992.
- The Management of Health and Safety at Work (Amendment) Regulations 1994.
- The Health Safety (Young Persons) Regulations 1997.
- Part III of the Fire Precautions (Workplace) Regulations 1997.

In addition to the above revocations, Regulation 29 amends a number of statutory instruments. These 'consequential amendments' are detailed in Schedule 2 to the Regulation. Sixteen statutory instruments are amended in total, these being:

- The Safety Representatives and Safety Committees Regulations 1977 (SI 1977/500).
- The Offshore Installations (Safety Representatives and Safety Committees) Regulations 1989 (SI 1989/971).
- The Railways (Safety Case) Regulations 1994 (SI 1994/237).
- The Suspension from Work (on Maternity Grounds) Order 1994 (SI 1994/2930).
- The Construction (Design and Management) Regulations 1994 (1994/3140).
- The Escape and Rescue from Mines Regulations 1995 (SI 1995/2870).
- The Main Miscellaneous Health and Safety Provisions Regulations 1995 (SI 1995/2005).
- The Quarries Miscellaneous Health and Safety Provisions Regulations 1995 (SI 1995/2036).
- The Borehole Sites and Operations Regulations 1995 (SI 1995/2038).
- The Gas Safety (Management) Regulations 1996 (SI 1996/551).

- The Health and Safety (Safety Signs and Signals) Regulations 1996 (SI 1996/341).
- The Health and Safety (Consultation with Employees) Regulations 1996 (SI 1996/1513).
- The Fire Precautions (Workplace) Regulations 1997 (SI 1997/1840).
- The Control of Lead at Work Regulations 1998 (SI 1998/543).
- The Working Time Regulations 1998 (SI 1998/1833).
- The Quarries Regulations 1999 (SI 1999/2024).

The amendments to these Regulations are limited to where specific reference is made to the Management of Health and Safety at Work Regulations 1992 (HSE, 1992). The amendments modify references to the MHSW Regulations 1992 with the insertion of corresponding reference to the Management of Health and Safety at Work Regulations 1999 (HSE, 1999). For example Regulation 16(1)(a) of the CDM Regulations 1994 gives reference to Regulation 9 of the MHSW 1992 – cooperation and coordination – this is now substituted with a reference to Regulation 11 of the MHSW 1999 – cooperation and coordination. For further specific detail of the amendments to the 16 statutory instruments the reader is directed to Schedule 2 accompanying the MHSW Regulations 1999.

### Regulation 30 – Transitional provision

This Regulation outlines that the substitution of these Regulations for the provisions of the Management of Health and Safety at Work Regulations 1992 'shall not affect the continuity of the law'.

## Undertaking risk assessment

The approved code of practice (HSE, 2000) highlights that a risk assessment should:

(a) ensure that the significant risks and hazards are addressed.
(b) ensure all aspects of the work actively are reviewed, including routine and non-routine activities. The assessment should cover all parts of the work activity, including those that are not under the immediate supervision of the employer, such as employees working off site as contractors, workers from one organization temporarily working for another organization, self-employed people, homeworkers and mobile employees.
(c) take account of non-routine operations, e.g. maintenance, cleaning operations, loading and unloading of vehicles, changes in production cycles, emergency response arrangements.

(d)  take account of the management of incidents such as interruptions to work activity, which frequently cause accidents, and consider what procedures should be followed to mitigate the effects of the incident.

(e)  be systematic in identifying hazards and looking at risks, whether one risk assessment covers the whole activity or the assessment is divided up. For example, it may be necessary to look at activities in groups such as machinery, transport, substances, electrical etc., or to divide the work site on a geographical basis. In other cases, an operation by operation approach may be needed, dealing with materials in production, dispatch, offices etc. The employer or self-employed person should always adopt a structured approach to risk assessment to ensure all significant risks or hazards are addressed. Whichever method is chosen, it should reflect the skills and abilities of the individuals carrying out that aspect of the assessment.

(f)  take account of the way in which work is organized, and the effects this can have on health.

(g)  take account of the risks to the public.

(h)  take account of the need to cover fire risks.

While it is essential that employers with five or more employees undertake to record the significant hazards identified by risk assessment, a single standard format for the recording of such risk assessment does not exist. It is common practice, though not required by law, to utilize a method statement approach throughout construction projects. The utilization of a health and safety method statement demands the evaluation of risk and the documentation and carrying out of an appropriately controlled approach to the considered construction process. It is not within the scope of this chapter to present a format for the recording of risk assessment and writing of method statements, however, this is covered in Chapters 17 and 18.

## Key points

This chapter has identified that:

- The Management of Health and Safety at Work Regulations 1992 were developed to implement the general provisions of the European Framework Directive (89/391/EEC). The 1992 Regulations were revoked and replaced by the Management of Health and Safety at Work Regulations 1999.
- The Regulations (HSE, 1999) consist of 30 Regulations and are supported by a Health and Safety Commission approved code of practice (HSE, 2000).
- Risk assessment is an important aspect of implementing the Regulations.

## References

European Commission (EC) (1989) *Directive 89/391/EEC on the introduction of measures to encourage improvements in the health and safety of workers at work*, HMSO, London.

Health and Safety Executive (HSE) (1974) *Health and Safety at Work, etc. Act 1974*, HMSO, London.

Health and Safety Executive (HSE) (1989) *Health and Safety Information for Employers 1989/2923*.

Health and Safety Executive (HSE) (1992) *Management of Health and Safety at Work Regulations 1992*, HMSO, London.

Health and Safety Executive (HSE) (1995) *Health and Safety Information for Employees (Modifications and Repeals) Regulations 1995*.

Health and Safety Executive (HSE) (1999) *Management of Health and Safety at Work Regulations 1999, SI 1999/3242*, HMSO, London.

Health and Safety Executive (HSE) (2000) *Management of Health and Safety at Work. Approved Code of Practice*, second edition, HMSO, London.

# 8 The Construction (Design and Management) Regulations 1994

## Introduction

The Construction (Design and Management) Regulations 1994 (CDM Regulations) are of significant importance to the health and safety management of UK construction projects. The Regulations emphasize the *management* of health and safety *throughout all stages* of construction projects and have brought about a shift in the approach to health and safety management within the UK construction industry. This shift has been towards *proactive* safety management throughout the processes of planning, design and production. In developing a proactive safety management approach, the Regulations place specific duties upon numerous parties to certain construction projects. Moreover, all parties to certain projects must now be coordinated and managed throughout all stages of the project. The Regulations also place responsibilities upon two new duty holders – the planning supervisor, appointed by the client and having overall responsibility for health and safety in design and planning, and the principal contractor, who must manage health and safety on site.

In presenting the CDM Regulations, this chapter outlines:

- The basis and structure of the Regulations.
- The application of the Regulations and exemptions to application.
- The key requirements and key features of CDM management.
- The CDM roles, duties and responsibilities of construction project participants.
- Issues appertaining to the appointment of the planning supervisor, consultants and the principal contractor.
- Aspects of enforcement of the Regulations and fines imposed by prosecution.

## Basis and structure of the Regulations

A European study of construction industry fatal accidents showed that although the primary cause of 37 per cent of the accidents were failures of the construction site management and workers, 28 per cent of accidents could be attributed to poor planning and 35 per cent due to unsafe design. An important conclusion was therefore that over 60 per cent of accidents were due to decisions made before the work began.

(Croner, 1994)

In recognition of both the unsatisfactory health and safety record of the European construction industry, and the need for greater consideration to be given to health and safety management throughout the whole construction process, the European Council issued Directive 92/57/EEC – The Temporary or Mobile Construction Sites Directive (European Commission, 1992).

In the UK this Directive was implemented by the CDM Regulations. The Regulations do not replace or supersede previous health and safety legislation, instead they further supplement the implementation of the Health and Safety at Work, etc. Act 1974 (HSE, 1974).

The CDM Regulations were made in December 1994, laid before Parliament in January 1995, and came into force on 31 March 1995. Transitional arrangements existed until 31 December 1995 for projects already in post-tender phase. As such the Regulations came into full force on 1 January 1996.

**Structure**  The CDM Regulations consist of 24 Regulations applicable to the development, design and production phases of a construction project. It is not the intention of this chapter to reproduce and provide commentary on each of the Regulations. Instead, where appropriate, it provides a number of definitions and explanations that will help place health and safety management in the context of the Regulations.

The Regulations are:

| | |
|---|---|
| Regulation 1 | Citation and commencement |
| Regulation 2 | Interpretation |
| Regulation 3 | Application of Regulations |
| Regulation 4 | Clients and agents of clients |
| Regulation 5 | Requirements on developer |
| Regulation 6 | Appointments of planning supervisor and principal contractor |
| Regulation 7 | Notification of projects |
| Regulation 8 | Competence of planning supervisor, designers and contractors |
| Regulation 9 | Provision for health and safety |
| Regulation 10 | Start of construction phase |
| Regulation 11 | Client to ensure information is available |
| Regulation 12 | Client to ensure health and safety file is available for inspection |
| Regulation 13 | Requirements on designer |

Regulation 14   Requirements on planning supervisor
Regulation 15   Requirements relating to the health and safety plan
Regulation 16   Requirements on and powers of principal contractor
Regulation 17   Information and training
Regulation 18   Advice from, and views of, persons at work
Regulation 19   Requirements and prohibitions on contractors
Regulation 20   Extension outside Great Britain
Regulation 21   Exclusion of civil liability
Regulation 22   Enforcement
Regulation 23   Transitional provisions
Regulation 24   Repeals, revocations and modifications

## Application of the Regulations

The Regulations apply to all construction projects with a few exceptions. The exceptions are work for a domestic client, projects which do not employ five or more people on site, and projects whose duration does not exceed 30 working days or 500 person hours. However, in the instances where the CDM Regulations do not apply other health and safety legislation may still apply.

The CDM Regulations apply to:

1. Any 'project' that involves 'construction work' that is 'notifiable' (these terms are described below).
2. Any project that is not notifiable but where five or more persons will be involved at any one time.

Definitions   **Project**
A project means 'a project which includes or is intended to include construction work' (HSE, 1994).

**Construction work**
Construction work means (Croner, 1994):

> the carrying out of any building, civil engineering or engineering construction work and includes any of the following:
> (a)  The construction, alteration, conversion, fitting out, commissioning, renovation, repair, upkeep, redecoration or other maintenance, de-commissioning, demolition or dismantling of a structure.
> (b)  The preparation for an intended structure, including site clearance, exploration, investigation (but not site survey) and excavation, and laying or installing the foundations of a structure.

(c)   The assembly . . . or . . . disassembly of a prefabricated structure.

(d)   The removal of a structure or waste resulting from demolition or dismantling of a structure.

(e)   The installation, commissioning, maintenance, repair, or removal of mechanical, electrical, gas, compressed air, hydraulic, telecommunications, computer or similar services which are normally fixed within or to a structure.

### Structure

Structure means (Croner, 1994):

(a)   Any building, steel or reinforced concrete structure . . . . railway . . . or . . . tramway line, dock, harbour, inland navigation (i.e. canal), tunnel, shaft, bridge, viaduct, waterworks, reservoir . . . pipeline . . . , cable, aqueduct, sewer, sewage works, gasholder, road, airfield, sea defence works, river works, drainage works, earthworks, lagoon, dam, wall, caisson, mast, tower, pylon, underground tank, earth retaining structure or structure designed to preserve or alter any natural feature.

(b)   Any formwork, falsework, scaffold or other temporary structure designed or used to provide support or means of access during construction work.

(c)   Any fixed plant from which a person can fall more than two metres during installation, commissioning, de-commissioning or dismantling work.

### Notifiable

A notifiable project is one where (HSE, 1994):

- the production stage on site will exceed 30 days, or
- the production stage on site will exceed 500 person days.

Where projects are notifiable, the planning supervisor must write to the HSE and provide the following information:

- date of notification
- construction project address
- name and address of client(s) or clients agent
- type of project, i.e. construction work type
- name and address of the planning supervisor
- signed declaration by the planning supervisor of appointment
- name and address of the principal contractor
- signed declaration by the principal contractor of appointment
- commencement date for start on site
- planned duration of the works

- estimated maximum number of persons to be at work on the site
- planned number of contractors to be at work on the site
- name and address of contractors already appointed.

**Exemptions to the application of the Regulations**

There are a number of situations where the Regulations do not apply. These include works undertaken by local authorities, domestic householders and projects not notifiable by that criteria specified above.

## Key requirements of the CDM Regulations

Two fundamental components of health and safety management have been introduced by the CDM Regulations. These new requirements of construction health and safety management are as follows:

1. The development of a health and safety plan. This is divided into two parts:
   (i)   the pre-tender health and safety plan
   (ii)  the construction phase health and safety plan.
2. The compilation of a health and safety file.

The development, structure, content and implementation of both the health and safety plan and the health and safety file is described in Chapter 17.

## Key features of CDM management

The key features of health and safety management within the CDM Regulations are:

- *Risk assessment* – parties must identify and assess project health and safety risk to comply with their duties.
- *Competence and adequate resources* – every designer, contractor and planning supervisor must be pre-qualified by assessment to ensure that they are competent and have the necessary resources to fulfil their duties for health and safety.
- *Cooperation and coordination* – all parties have an obligation to cooperate and coordinate their efforts to identify and mitigate health and safety risk.
- *Provision of information* – all parties have a duty to share information pertinent to health and safety, for the development of the project's health and safety plan and health and safety file.

## Roles, duties and responsibilities of project participants

The CDM Regulations place a duty upon clients, consultants and contractors to coherently and methodically think about their contributions to project health and safety. Moreover, their inputs must be coordinated and managed throughout all phases of the construction process. Figure 8.1 outlines those parties with CDM health and safety involvement throughout the differing phases of a construction project.

**The client**  The client may be defined as 'any person for whom a project is carried out'. The client may appoint an agent to act on their behalf, and to all intents and purposes under the Regulations the agent is treated as the client. An agent must be assessed as competent and have the necessary resources to carry out the duties of the client they represent. When an agent is appointed, this arrangement must be notified in writing to the HSE.

The client's key duties are:

- To determine if the project must comply with the CDM Regulations.
- To appoint a *planning supervisor* and *principal contractor* for the project

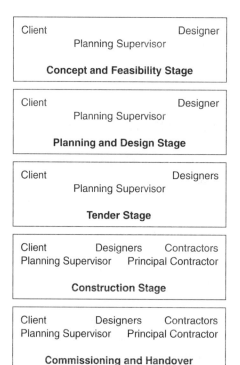

**Fig. 8.1** Parties with CDM health and safety responsibilities throughout a construction project

Client                                Designer
         Planning Supervisor
**Concept and Feasibility Stage**

Client                                Designer
         Planning Supervisor
**Planning and Design Stage**

Client                                Designers
         Planning Supervisor
**Tender Stage**

Client            Designers        Contractors
Planning Supervisor    Principal Contractor
**Construction Stage**

Client            Designers        Contractors
Planning Supervisor    Principal Contractor
**Commissioning and Handover**

(these parties are described subsequently) and ensure that both are competent and have adequate resources to fulfil their obligations.

- To ensure that any design consultants and contractors that they nominate are competent and have adequate resources.
- To provide the planning supervisor with information pertinent to health and safety matters on the project.
- To ensure that the construction work on site does not commence until an appropriate health and safety plan has been prepared.
- To ensure that the project's health and safety file is available for perusal following completion of the project.

**The planning supervisor**

The planning supervisor may be defined as 'a person appointed by the client who has overall responsibility for coordinating the health and safety aspects of the design and planning phase.' The planning supervisor is a role created by the Regulations. The planning supervisor may be either an individual or a multidisciplinary team. Where a project is large and complex it is not unusual for a planning supervisory team approach to be adopted. A number of organizations now offer such a consultative facility. Irrespective of who the client employs to fulfil this role, it is the client's responsibility to ensure that the appointee is competent and has adequate resources.

There must be a planning supervisor in place throughout the duration of the project, although the appointee can be changed at the discretion of the client as the project proceeds. The appointed person(s) and any changes made must be notified in writing to the HSE.

The planning supervisor's key duties are:

- To ensure that designers comply with their duties, in particular the identification, avoidance and mitigation of risk and also in providing adequate information on health and safety for construction, maintenance and demolition.
- To encourage cooperation between the designer and other designers and/or other consultants for the purposes of promoting health and safety.
- To ensure that a health and safety plan is prepared before the principal contractor is appointed.
- To give advice, if requested, to the client on the appointment of designers and contractors, and to advise contractors appointing designers.
- To advise the client on the health and safety plan before construction work commences.
- To notify the HSE of health and safety management arrangements for the project.
- To ensure that the health and safety file is complete and passed to the client at the end of the project.

**The designer**  The designer may be defined as 'the person(s) who prepare designs (or engineers) or arrange for persons under their control to develop a design.' The designer should ensure that the health and safety of those persons involved in construction work, maintenance and repair are considered during the formulation of design to reduce aspects of risk on the project.

The designer's key duties are:

- To ensure that the client is aware of their duties under the Regulations.
- To develop designs which avoid risks to health and safety.
- To consider, when formulating designs, risks which may arise during construction or maintenance.
- To reduce the level of risk where it is not possible to eliminate the risk.
- To consider measures which will protect workers where avoidance or reduction of risk to a safe level is not possible.
- To ensure that the design includes adequate information on health and safety and to pass this information on to the planning supervisor for inclusion in the health and safety plan.
- To cooperate with the planning supervisor and, where necessary, other designers to enable them to comply with health and safety legislation.

**The principal contractor**  Principal contractors may be defined as 'persons who undertake or manage construction work or who arrange for persons under their control to undertake or manage construction work'. The client has a duty to appoint a principal contractor for the project as soon as it is practicable to do so. The client can be the principal contractor providing that the client is, by definition, a contractor.

It is the responsibility of the principal contractor to take over and develop the health and safety plan initiated by the planning supervisor. The principal contractor is also responsible for coordinating health and safety activities of all contractors on site.

The principal contractor's key duties are:

- To develop and implement the construction health and safety plan.
- To arrange for competent and adequately resourced contractors to undertake work where it is subcontracted.
- To ensure the coordination and cooperation of all contractors.
- To determine from contractors the findings of their risk assessments and details of how they propose to carry out high-risk operations.
- To provide information to contractors of project site risks.
- To ensure that contractors and operatives comply with site rules and procedures specified in the health and safety plan.
- To monitor health and safety performance throughout the project.
- To ensure that all operatives are fully informed and consulted on health and safety matters.

- To allow only authorized persons onto the site.
- To display in a prominent position the notification of the project to HSE.
- To ensure that information is passed to the planning supervisor for inclusion in the health and safety file.

**Contractors**    Contractors may be defined as 'persons who undertake or manage construction work or who arrange for persons under their control to undertake or manage construction work as a subcontractor of the principal contractor.' Contractors include those persons who are self-employed. Contractors are required to cooperate with the principal contractor and provide information concerning health and safety risk assessment and management procedures.

The contractor's key duties are:

- To provide information for the health and safety plan concerning the risks arising from their work together with details of the actions they will take to manage those risks.
- To undertake their work while complying with directions from the principal contractor and with rules and procedures specified in the health and safety plan.
- To provide information for the health and safety file and report to the HSE under the Reporting of Injuries, Diseases and Dangerous Occurrences Regulations 1995 (RIDDOR) (HSE, 1995a) instances of dangerous occurrences, ill health and injuries.
- To provide adequate information to their employees on health and safety matters.

All contractors must provide their employees with:

- the name of the planning supervisor
- the name of the principal contractor
- details of the health and safety plan

## Appointment of the planning supervisor, consultants and the principal contractor

The CDM Regulations bring the health and safety roles, duties and responsibilities of all parties involved in a project sharply into focus. Previous to the CDM Regulations there was no expressed need to consider health and safety when determining roles, appointments and contracts. In setting out the terms agreed between parties to a project, contractual arrangements for projects subject to the CDM Regulations should now express health and safety management responsibilities.

## Appointment of the planning supervisor

The CDM Regulations require that the planning supervisor be appointed by the client as soon as is practicable so that the client can be aided in the fulfilment of his/her duties. It is the client's responsibility to monitor and review progress of the planning supervisor's duties as the appointment may need to be terminated and a new appointment made should there be any lapse in competence or failure to resource the task adequately.

In some instances a change of planning supervisor may be deliberately scheduled to take place during the project. In design-build procurement for example, the planning supervisor may be appointed to deliver services during the conceptual design stage and a different planning supervisor appointed at the detailed design stage.

The CDM Regulations require the planning supervisor to ensure that a pre-tender health and safety plan is compiled. It is not necessarily the responsibility of the planning supervisor to prepare the plan her or himself. If the client wishes the planning supervisor to specifically prepare the plan, then the client would need to specify this. Similarly, the planning supervisor has to ensure that the health and safety file is compiled, but there is no responsibility, unless specifically determined, for the planning supervisor to compile it. There is, therefore, a clear need for both the client and planning supervisors to consider carefully and express what duties are to be performed both generally under the Regulations and in meeting any additional requirements of the client.

The Royal Institute of British Architects (RIBA) has available to clients a Form of Appointment as Planning Supervisor (RIBA, 1995a) for use when arranging the services of construction professionals as planning supervisors. The form consists of a memorandum of agreement between the client and planning supervisor setting out a schedule of services to be provided by the planning supervisor.

## Appointment of consultants

The CDM responsibilities of design consultants apply to all projects irrespective of the size and type of project or the numbers of staff employed. Whereas health and safety management responsibilities of other parties to a project are determined by whether the project is CDM applicable or not, this is not the case for designers. The responsibility of designers to design out risk is applicable to all undertakings as an exercise of their profession.

The RIBA has a Standard Form of Agreement for the Appointment of an Architect (RIBA, 1992) that incorporates a supplement to include the application of the CDM Regulations. The document sets out conditions of appointment and services expected to meet with health and safety requirements.

The RIBA document, Conditions of Engagement for the Appointment of an Architect (RIBA, 1995b) presents guidance on the responsibilities of the architect when acting only as the designer, i.e. when a separate planning supervisor is appointed and the designer does not act in that capacity.

**Appointment of the principal contractor**

The appointment of the principal contractor is a significant element in ensuring that health and safety is effectively managed during the construction phase. The client should ensure that any prospective principal contractor is competent, has considered health and safety in their tender and included for health and safety management in their tender bid price. A client may take measures to ensure this, by first adopting a health and safety pre-qualification assessment of all prospective principal contractors, which may take the form of a questionnaire, and second by inviting a detailed health and safety policy to be submitted for consideration.

There is no single set procedure that must be followed for health and safety during the tender stage. However, the HSE publication *A Guide to Managing Health and Safety in Construction* (HSE, 1995b) is available as a source of advice to clients.

From the client's perspective, several key points should be noted:

- The tender documents which are sent to prospective principal contractors should be formatted in such a way that the methods by which the principal contractor includes for and prices health and safety management should be transparent.
- The tender documents should include a pre-tender health and safety plan that will inform the prospective principal contractor of significant risks which have been identified and which should be considered in the tender in terms of management requirements and cost implications.
- The tendering period should be sufficient to allow prospective principal contractors to comprehensively consider the health and safety management aspects and tender appropriately.

From the prospective principal contractor's perspective, they should be able to demonstrate that they:

- are competent to manage health and safety on the project. This can be achieved by providing records from previous projects undertaken and profiles of qualified staff to be employed on the project.
- have the adequate resources to manage the health and safety matters which are identified. This can be demonstrated in the costings provided in the tender price.
- have implemented a health and safety policy within their organization which is applied to their projects. This should be available in existing company documentation such as statements of annual reports.
- have formal procedures for managing those health and safety aspects identified in the pre-tender health and safety plan.

In considering and evaluating submitted tenders the client may take advice from the planning supervisor. In addition, the client and planning supervisor may meet with prospective principal contractors to discuss services to the contract in further detail.

## Enforcement of the CDM Regulations

Breach of the Regulations is a criminal offence and as such can result in prosecution. The penalties resulting from a successful prosecution are outlined in Chapter 6. The bringing of a civil action as a result of a breach of the CDM Regulations is in the main forbidden by Section 47 of the Health and Safety at Work, etc. Act 1974. Civil actions can only be brought for breach of Regulations 10 and 16(1)(c). Regulation 10 is of concern to the client and Regulation 16(1)(c) is of concern to the planning supervisor:

### Regulation 10

Start of construction phase –
Every client shall ensure, so far as is reasonably practicable, that the construction phase of any project does not start unless a health and safety plan complying with Regulation 15(4) has been prepared in respect of that project.

### Regulation 16(1)

The principal contractor appointed for any project shall –
(c) take reasonable steps to ensure that only authorized persons are allowed into any premises or part of premises where construction work is being carried out.

Around 90 per cent of HSE prosecutions are successful, though it is commonly considered that the bringing of such prosecution is undertaken as a last resort. Of these prosecutions only a very few are brought under the CDM Regulations. Health and safety prosecutions are more commonly brought under the Health and Safety at Work, etc. Act 1974.

## CDM fines

This section outlines some fines imposed under the CDM Regulations. It is worth noting that the majority of successful prosecutions brought under the Regulations have been against the client.

- A client failed to appoint a planning supervisor. The breach was of Regulation 6(1). The fine was £10,000.

- A principal contractor failed to adequately assess the competence of a subcontractor. The breach was of Regulation 8(3). The fine was £3,000.
- A client allowed the construction phase of a project to proceed without an adequate health and safety plan. The breach was of Regulation 10. The fine was £2,000.
- A client failed to provide information about the condition of the project premises. The breach was of Regulation 11. The fine was £2,000.
- An architectural practice failed to inform the client of his duties under the CDM Regulations. The breach was of Regulation 13(1). The fine was £500.
- A planning supervisor failed to provide adequate advice to the client. This was in breach of Regulation 14(c). The fine was £15,000 with £5,000 costs.
- A principal contractor failed to have an adequate health and safety plan in operation. During the project one person was hospitalized for six weeks and another killed. The breach was of Regulation 15(4). The fine was £20,000 with £3,000 costs.
- A principal contractor failed to provide a construction phase health and safety plan and also failed to ensure the adequate training of the employees of a subcontractor. The breaches were of Regulations 15(4)(a) and 17(2)(b). The fines were £2,500 and £1,500 respectively.

## Further information on health and safety management

The approved code of practice that accompanies the Regulations gives practical guidance regarding the Regulations and the management of construction for health and safety. For further information about the Regulations the reader is directed to this approved code of practice (HSC, 1995b) and to other HSC publications, such as *Designing for Health and Safety in Construction* (HSC, 1995a).

## Key points

This chapter has identified that:

- The Construction (Design and Management) Regulations 1994 (HSE, 1994) is the most significant piece of legislation influencing the management of health and safety in construction.
- The Regulations place specific duties upon contractual parties to a construction project.
- Health and safety must be coordinated and managed by the parties holistically throughout all stages of the project.

## References

Croner (1994) *Croner's Management of Construction Safety*, Croner Publications Ltd., Kingston upon Thames, Surrey.

European Commission (EC) (1992) *The Temporary or Mobile Construction Sites Directive*, HMSO, London.

Health and Safety Commission (HSC) (1995a) *Designing for Health and Safety in Construction*, HMSO, London.

Health and Safety Commission (HSC) (1995b) *Managing Construction for Health and Safety*, HMSO, London.

Health and Safety Executive (HSE) (1974) *The Health and Safety at Work, etc. Act 1974*, HMSO, London.

Health and Safety Executive (HSE) (1994) *The Construction (Design and Management) Regulations 1994*, HMSO, London.

Health and Safety Executive (HSE) (1995a) *The Reporting of Injuries, Diseases and Dangerous Occurrences Regulations 1995 (RIDDOR)*, HMSO, London.

Health and Safety Executive (HSE) (1995b) *A Guide to Managing Health and Safety in Construction*, HMSO, London.

Health and Safety Executive (HSE) (1999) *The Management of Health and Safety at Work Regulations 1999* HMSO, London.

Royal Institute of British Architects (RIBA) (1992) *Standard Form of Agreement for the Appointment of an Architect (SFA/92)*, RIBA, London.

Royal Institute of British Architects (RIBA) (1995a) *Form of Appointment as Planning Supervisor (PS/45)*, RIBA, London.

Royal Institute of British Architects (RIBA) (1995b) *Conditions of Engagement for the Appointment of an Architect (CE/95)*, RIBA, London.

# 9 The Construction (Health, Safety and Welfare) Regulations 1996 and associated welfare legislation

## Introduction

This chapter looks at aspects of construction industry welfare legislation. In doing so it provides an insight into the following Regulations:

The Construction (Health, Safety and Welfare) Regulations 1996 (CHSWR).
The Reporting of Injuries, Diseases and Dangerous Occurrences Regulations 1995 (RIDDOR).
The Control of Substances Hazardous to Health (Amendment) Regulations 1999 (COSHH).
The Personal Protective Equipment at Work Regulations 1992 (PPE).

More specifically, this chapter presents:

- The basis, structure and outline of the CHSW Regulations.
- The duties under RIDDOR to report, record and investigate accidents.
- The duties imposed upon employers under the COSHH Regulations and an outline definition of 'hazardous substances'.
- A definition of personal protective equipment and an introduction to the Personal Protective Equipment at Work Regulations.

## The Construction (Health, Safety and Welfare) Regulations 1996

The Construction (Health, Safety and Welfare) Regulations 1996 (HSE, 1996a) came into force on 2 September 1996 and revoked the Construction (Health and Welfare) Regulations 1966 (HSE, 1966a) and the Construction (Working Places) Regulations 1966 (HSE, 1966b).

The Regulations apply to all construction work and Regulation 2 defines construction work as:

the carrying out of any building, civil engineering or engineering construction work and includes any of the following:

(a) The construction, alteration, conversion, fitting out, commissioning, renovation, repair, upkeep, redecoration or other maintenance, de-commissioning, demolition or dismantling of a structure.

(b) The preparation for an intended structure, including site clearance, exploration, investigation (but not site survey) and excavation, and laying or installing the foundations of a structure.

(c) The assembly . . . or . . . disassembly of a prefabricated structure.

(d) The removal of a structure or waste resulting from demolition or dismantling of a structure.

(e) The installation, commissioning, maintenance, repair, or removal of mechanical, electrical, gas, compressed air, hydraulic, telecommunications, computer or similar services which are normally fixed within or to a structure.

The Regulations do not apply to locations where construction work as defined above does not take place. For such locations, including site offices, the Workplace (Health, Safety and Welfare) Regulations 1992 (HSE, 1992a) apply.

The Construction (Health, Safety and Welfare) Regulations 1996 consist of 35 Regulations and 10 appended Schedules. This chapter outlines Regulations 3–30. Regulations 1 and 2 concern 'citation and commencement' and 'interpretation'; for additional detail the reader is referred to the Regulations and their approved code of practice (HSE, 1996b).

### Regulation 3 – Application

The Regulations only apply to construction work carried out by a person at work, they do not apply to DIY work. The Regulations also do not apply to locations on site set aside for purposes other than construction work.

### Regulation 4 – Persons upon whom duties are placed by these Regulations

This Regulation states that those persons upon whom duties are placed are employers, self-employed persons and employees.

It is the duty of every person at work to cooperate with any other persons with a duty imposed by the Regulations. It is also the duty of every person working under the control of another person to report to that person any defect that may endanger a person's health and safety.

### Regulation 5 – Safe places of work

Construction sites must provide, so far as is reasonably practicable, safe and appropriate access and egress, sufficient working space and be made suitably safe.

### Regulation 6 – Falls

Suitable and sufficient steps are to be taken, as far as is reasonably practicable, to prevent persons falling. Measures taken are to include toe-boards, guard rails, barriers or working platforms. The Schedules to the Regulations specify the requirements in detail. With regard to lifting equipment and operations The Lifting Operations and Lifting Equipment Regulations 1998 (LOLER) replace paragraph 3 of Schedule 9 of the Regulations.

### Regulation 7 – Fragile material
Suitable and sufficient steps are to be taken to prevent any person falling through fragile material.

### Regulation 8 – Falling objects
Suitable and sufficient steps are to be taken, as far as is reasonably practicable, to prevent material or objects from falling. Measures taken are to include toe-boards, guard rails, barriers or working platforms. If this is not reasonably practicable then measures are to be taken to prevent injury to persons as a result of falling objects.

### Regulation 9 – Stability of structures
All practicable steps are to be taken to prevent existing or new structures from weakening and collapsing as a result of construction work. Structural supports are also to be erected and dismantled only under the supervision of a competent person.

### Regulation 10 – Demolition or dismantling
So far as is reasonably practicable, appropriate measures are to be taken to prevent injury as a result of demolition or dismantling. Supervision of the activity is to be undertaken by a competent person.

### Regulation 11 – Explosives
Suitable and sufficient steps are to be taken to prevent risk of injury from an explosion or associated flying material.

### Regulation 12 – Excavations
All practicable steps are to be taken to prevent danger of accidental collapse of excavations. Measures are also to be taken, so far as is reasonably practicable and under the supervision of a competent person, to prevent persons being trapped or buried by material that has fallen or been dislodged. Appropriate measures must also be taken to prevent persons, vehicles, plant, equipment or materials from falling into an excavation. Excavation work shall not begin without the identification of services and, so far as is reasonably practicable, the prevention of injury from such services.

### Regulation 13 – Cofferdams and caissons
These must be suitable for their purpose in design, construction and materials and must be properly maintained.

### Regulation 14 – Prevention of drowning

Wherever appropriate suitable and sufficient steps are to be taken, so far as is reasonably practicable, to prevent drowning. Suitable rescue equipment is to be provided whenever a risk of drowning in liquid exists.

### Regulation 15 – Traffic routes

So far as is reasonably practicable, pedestrians and vehicles on construction sites are to be organized so as to be safe. Traffic routes are to be appropriate in size and location and sufficient space allowed between pedestrian doors and gates and traffic routes to provide for adequate visibility. Also, each loading bay must have at least one pedestrian-only exit.

### Regulation 16 – Doors and gates

Doors and gates must be fitted with suitable safety devices to prevent injury.

### Regulation 17 – Vehicles

Provides for the safe operation of vehicles.

### Regulation 18 – Prevention of risk from fire, etc.

Suitable and sufficient steps are to be taken, so far as is reasonably practicable, to prevent risks of injury from fire, flood, explosion or asphyxiating substances.

### Regulation 19 – Emergency routes and exits

Sufficient emergency routes that are clear of obstructions and suitably signed are to be provided. Where necessary emergency lighting is to be provided. Paragraph (4) of the Regulation provides that any emergency routes and exits provided to enable any person on a construction site to 'reach a place of safety quickly in the event of danger' shall have regard to:

(a) the type of work for which the construction site is being used
(b) the characteristics and size of the construction site and the number and location of places of work on that site
(c) the plant and equipment being used
(d) the number of persons likely to be present on the site at any one time; and
(e) the physical and chemical properties of any substance or materials on or likely to be on site.

### Regulation 20 – Emergency procedures

Provides that employers ensure suitable and sufficient emergency arrangements for all foreseeable emergencies. Evacuation procedures must be undertaken with regard to paragraph (4) of Regulation 19. Arrangements for dealing with any foreseeable emergency shall:

(2)(b)  designate an adequate number of persons who will implement the arrangements; and

(c)  include any necessary contacts with external emergency services, particularly as regards rescue work and fire-fighting.

The insertion of (b) and (c) into paragraph (2) was further to the Management of Health and Safety at Work Regulations 1999 (HSE, 1999a). Paragraph 3 of the Regulation also provides sufficient steps be taken to ensure that:

(3)(a)  every person to whom the arrangements extend is familiar with those arrangements; and

(b)  the arrangements are tested by being put into effect at suitable intervals.

### Regulation 21 – Fire detection and fire-fighting

Fire-fighting equipment, alarms and fire detectors must be suitably located on site and be properly maintained and tested. So far as is reasonably practicable every person on site is to be trained in the use of such equipment.

### Regulation 22 – Welfare facilities

Those persons in control of a construction site are duty bound to ensure compliance with these Regulations on that site. Welfare provisions are to be ensured by employers and self-employed. These provisions include: sufficient sanitary conveniences in accordance with Schedule 6; adequate washing facilities – including showers when required – in accordance with Schedule 6; an adequate and accessible supply of drinking water; accommodation for clothing storage and the drying of clothing; accommodation for the changing of clothing when special clothing is required for work; and, so far as is reasonably practicable, facilities for rest with separate non-smoking areas and facilities for preparing and eating meals.

### Regulation 23 – Fresh air

Provides that, so far as is reasonably practicable, sites be properly ventilated with a supply of fresh or purified air. Plant utilized to deliver air must be fitted with a visible or audible device warning of defect.

### Regulation 24 – Temperature and weather protection

During work indoors, so far as is reasonably practicable, the temperature must be reasonable. During work outdoors, so far as is reasonably practicable, suitable protective clothing and equipment is to be provided.

### Regulation 25 – Lighting

Provides for suitable and sufficient lighting that does not affect the reading of signs and the like. Where failure of artificial lighting would result in a risk to health and safety necessary emergency lighting is to be installed.

### Regulation 26 – Good order

So far as is reasonably practicable construction sites are to be kept clean and in good order.

### Regulation 27 – Plant and equipment

This Regulation has now been revoked by Provision and Use of Work Equipment Regulations 1998 (PUWER), SI 1998/2306 (HSE, 1998).

### Regulation 28 – Training

Adequate safety information is to be given to each person and each person is to be adequately trained or supervised appropriately to prevent risks of injury.

### Regulation 29 – Inspection

Schedule 7 to the Regulations identifies those works requiring periodic inspection, these include cofferdams and caissons, excavation areas, working platforms and suspension equipment.

### Regulation 30 – Reports

A report is to be prepared for every inspection made under Regulation 29. This report is to be retained for a period of three months after completion.

### Regulations 31–35

Concern exemption certificates, extension outside Great Britain, enforcement in respect of fire, modifications and revocations. For additional detail of the Regulations the reader is referred to the approved code of practice (HSE, 1996b).

## The Reporting of Injuries, Diseases and Dangerous Occurrences Regulations 1995 (RIDDOR)

The Reporting of Injuries, Diseases and Dangerous Occurrences Regulations 1995 (HSE, 1995), often referred to as RIDDOR, came into force 1 April 1996. The Regulations require that certain construction site accidents must be reported to the HSE. The occurrences that require reporting are as follows.

- Fatal and serious accidents.
- Less serious accidents where a person is unfit for work for more than three consecutive days.
- A dangerous incident where persons are placed at risk.
- A specified disease associated with a person's job.

**Reporting mechanism**  It is the principal contractor's responsibility to inform HSE of the occurrence of a reportable incident. The HSE provides a series of report forms for this purpose. HSE report forms must be completed and returned to HSE within specified time periods according to the nature of the incident. For example, a serious injury to a person must be reported immediately by telephone and confirmed in writing within ten days.

A contractor must inform the principal contractor of any incidents occurring to their employees and these must be reported to HSE in the prescribed way.

**Record keeping**  Records of any reportable occurrence must be kept. The main information recorded is: the date, time and place of the occurrence; details of the injured or unwell person(s); a brief description of the incident or disease; and reference to the initial reporting mechanism, where for example a written report is followed a verbal report to HSE. In addition to record keeping required by RIDDOR, an employer is bound under the Social Security (Claims and Payments) Regulations 1979, SI 1979/628, to maintain an accident book. All incidents must be recorded in the book and the record must be retained for three years. The book must be available to audit.

**Investigation of an occurrence**  Any injury, disease and dangerous occurrence should always be investigated. The principal contractor should maintain a mechanism within its health and safety management system to facilitate investigation.

The HSE is empowered by law to investigate. HSE can request all health and safety documentation from the principal contractor and contractors on a project, gather information from workers and examine the site for evidence. The principal contractor is obliged to render whatever help is deemed necessary by the HSE inspectors.

Safety representatives within the principal contracting organization, appointed to represent employees, may also investigate any incident occurring on site. This investigation is for the purposes of identifying the cause and considering revised working methods. They may also investigate complaints raised by employees relating to health, safety and welfare matters. As with HSE investigations, all records, documentation and evidence must be available for scrutiny.

## The Control of Substances Hazardous to Health (Amendment) Regulations 1999 (COSHH)

The Control of Substances Hazardous to Health Regulations were first introduced in 1988. The 1988 Regulations were replaced in 1994 and again in 1999.

On 25 March 1999 the new Control of Substances Hazardous to Health Regulations 1999 were introduced (HSE, 1999b).

The COSHH Regulations impose duties upon employers to:

- *assess* health risks in any instances that employees could potentially be exposed to hazardous substances
- *prevent* (or *control*) exposure to hazardous substances
- *take reasonable action* to ensure measures that control exposure are undertaken appropriately
- *monitor*, wherever appropriate, employees' exposure to hazardous substances
- ensure appropriate *training and information* is given to persons who may be exposed to hazardous substances.

The Regulations apply to all substances that are capable of causing adverse health effects (Croner, 1997). Substances hazardous to health are defined as:

1. A substance listed in part 1 of the approved supply list as dangerous within the meaning of The Chemicals (Hazard Information and Packaging) Regulations 1993, SI 1997/1460, and for which an indication of danger specified for the substance on part V of that list is very toxic, toxic, harmful, corrosive or irritant.
2. A substance for which a maximum exposure limit is specified in the HSE's publication *EH40* (HSE, 1999c).
3. A biological agent which is capable of causing an infection, allergy, toxicity or other such hazard to the health of any person.
4. Dust of any kind, when substantially present in air.
5. A substance (other than those listed above in 1–4) which creates a hazard to the health of any person comparable to the hazards of those substances listed in 1–4 above.

Previously point 2 above outlined 'maximum exposure limits' contained in Schedule 1 of the Regulations. This Schedule has been removed in the 1999 Regulations with maximum exposure limits now identified within the Health and Safety Executive's publication *EH40: Occupational Exposure Limits* (HSE, 1999c).

The main changes brought about by the 1999 Regulations are:

1. The Health and Safety Commission's approval of maximum exposure limits in respect of specified substances – publication *EH40*.
2. The addition of definitions for 'respirable dust' and 'total inhalable dust'.
3. The duty of employers to provide personal protective equipment in accordance with The Personal Protective Equipment at Work Regulations 1992 (HSE, 1992).

**Breach of the Regulations**

Breach of the Regulations may result in criminal prosecution. If a statutory duty is breached a civil action may arise, as in the cases of *Bilton* v. *Fastnet Highlands Ltd* (1997), *Ogden* v. *Airedale Health Authority* (1996) and *Williams* v. *Farne Salmon & Trout Ltd* (1988).

## The Personal Protective Equipment at Work Regulations 1992 (PPE)

> Even where engineering controls and safe systems of work have been applied, some hazards might remain. These include injuries to lungs, e.g. from breathing in contaminated air; the head and feet, e.g. from falling materials; the eyes, e.g. from flying particles or splashes of corrosive liquids; the skin, e.g. from contact with corrosive materials; the body, e.g. from extremes of heat or cold. Personal protective equipment (PPE) is needed in these cases to reduce the risk.
>
> (HSE, 1997)

The Personal Protective Equipment at Work Regulations 1992 (HSE, 1992b) came into effect on 1 January 1993 and were developed to implement European Directive 89/656/EEC – to introduce minimum health and safety requirements for workers using personal protective equipment (PPE) at the workplace.

Where risks to personal health and safety are not sufficiently controlled, employers are duty bound by the Regulations to provide employees exposed to health and safety risks with suitable personal protective equipment. Self-employed persons are duty bound to ensure that they themselves are provided for.

Personal protective clothing is defined by Regulation (2) as all equipment and clothing which is intended to be worn or held by a person at work and which affords protection against one or more risks to health and safety. Clothing which is designed to protect against adverse weather conditions is also included in this definition. Personal protective equipment may be utilized to protect the eyes, head and neck, ears, hands, arms, feet, legs, the lungs and indeed the whole body. By its very nature personal protective equipment protects only the person wearing it and not those within the surrounding vicinity. Therefore, PPE should be used as a last resort in managing health and safety risk in the workplace.

Where risks to personal health and safety are not sufficiently controlled and PPE is utilized, the HSE (1997) advises employers and the self-employed to:

- Choose good quality products made to a recognized standard.
- Choose equipment which suits the wearer – in doing so, consider size, fit and weight.

- Ensure that equipment fits properly.
- Make sure that if more than one item of PPE is being worn they can be worn together.
- Instruct and train people in the use of equipment.

Such equipment could include, but is not limited to, eye protectors, gloves, safety footwear, safety helmets and disposable respirators.

Personal protective safety equipment must be of a suitable standard and must comply with the Personal Protective Equipment at Work Regulations 1992 (HSE, 1992b) which were drawn up to implement the Personal Protective Equipment Directive 89/686/EEC. The Directive provides for Community-wide standards as regards safety equipment. As such PPE must carry either the 'CE' mark or a declaration of conformity from the employer.

## Key points

This chapter has identified that:

- A number of significant pieces of legislation influence construction, these are:
  - The Construction (Health, Safety and Welfare) Regulations 1996 (HSE, 1996) (CHSW).
  - The Reporting of Injuries, Diseases and Dangerous Occurrences Regulations 1995 (HSE, 1995) (RIDDOR).
  - The Control of Substances Hazardous to Health (Amendment) Regulations 1999 (HSE, 1999b) (COSHH).
  - The Personal Protective Equipment at Work Regulations 1992 (HSE, 1992b) (PPE).

## References

Croner (1997) *Croner's Risk Assessment*, Croner Publications Ltd., Kingston upon Thames, Surrey.

Health and Safety Executive (HSE) (1966a) *The Construction (Health and Welfare) Regulations 1966*, HMSO, London.

Health and Safety Executive (HSE) (1966b) *The Construction (Working Places) Regulations 1966*, HMSO, London.

Health and Safety Executive (HSE) (1992a) *The Workplace (Health, Safety and Welfare) Regulations 1992*, HMSO, London.

Health and Safety Executive (HSE) (1992b) *The Personal Protective Equipment at Work Regulations 1992 (PPE)*, HMSO, London.

Health and Safety Executive (HSE) (1995) *The Reporting of Injuries, Diseases and Dangerous Occurrences Regulations 1995 (RIDDOR)*, HMSO, London.

Health and Safety Executive (HSE) (1996a) *Construction (Health, Safety and Welfare) Regulations 1996*, HMSO, London.

Health and Safety Executive (HSE) (1996b) *Construction (Health, Safety and Welfare) Regulations 1996. Approved Code of Practice*, HMSO, London.

Health and Safety Executive (HSE) (1997) *Essentials of Health and Safety at Work*, HMSO, London.

Health and Safety Executive (HSE) (1998) *The Provision and Use of Work Equipment Regulations 1998*, HMSO, London.

Health and Safety Executive (HSE) (1999a) *The Management of Health and Safety at Work Regulations 1999, SI 1999/3242*, HMSO, London.

Health and Safety Executive (HSE) (1999b) *Control of Substances Hazardous to Health (General ACOP), Control of Carcinogenic Substances (Carcinogens ACOP), and Control of Biological Agents (Biological Agents ACOP)*, HMSO, London.

Health and Safety Executive (HSE) (1999c) *EH40: Occupational Exposure Limits*, HSE, Sudbury, Suffolk.

# PART C  Effective health and safety management

# 10 Health and safety management within the construction process

## Introduction

This brief chapter introduces the conceptual thrust of health and safety management in the context of its application within the construction process. It should be seen as an introduction to the chapters which form Part C of this book. In appreciating the role, duties and responsibilities of the principal contractor, which is the focus of this book, it is important also to be cognizant of the contributions and responsibilities of other participants to the construction process.

The management of health and safety is without doubt one of the most important functions within and throughout the construction process. Construction work is intrinsically dangerous. Injury to persons on and around construction sites occur regularly. It is fortunate that many injuries are minor, but, others are serious and some are fatal. It was seen in Chapter 2 that the construction industry has over the last 20 years suffered a poor health and safety record. While the number of fatalities declined in the 1990s, this must not encourage complacency. Construction management has an unquestionable and definite challenge to ensure a safe working environment.

The Construction (Design and Management) Regulations 1994 (HSE, 1994), introduced welcome and much needed legislation to construction health and safety. The Regulations focus purposefully on the management of health and safety throughout the total construction process. Responsibility is unambiguous and specifically placed upon clients, designers and contractors to be proactive in the planning, coordination and management of health and safety. The Regulations bring into focus the identification of potential hazards to health and dangers to safety through each major phase of the construction process, together with the assessment of their risk.

## The health and safety plan

The CDM Regulations require that project health and safety planning and management be considered in two parts. The first part focuses on the client's project evaluation and design processes with the objective of producing a pre-tender health and safety plan. The second part focuses on the site production processes with the objective for the appointed principal contractor being to produce a construction phase health and safety plan. It is the essential element of planning within each part which forms the basis for a systematic management approach, within which risk assessment is an important theme.

Establishing effective health and safety management is the goal of the principal contractor supported by the inputs of the main parties to the construction project. The lead consultant, representing the client and working with sub-consultants, is charged with delivering the pre-tender health and safety plan and implementing project supervisory procedures that make a full contribution to project health and safety. The principal contractor is charged with delivering the construction health and safety plan. Moreover, the principal contractor must establish a management system and working procedures which ensure the maintenance of safe working conditions and practices.

A well-formulated project health and safety management approach will identify, assess and control risk both within and across the professional boundaries of the parties. Feedback loops within both the designer's and the contractor's systems will do much to ensure that information is directed not only within the span of control of the individual party but contributes to the management processes within other project phases. Within the context of the CDM Regulations a full and effective contribution will be in evidence as the principal output from the planning and management approach to health and safety in the delivery of the *health and safety file* – a complete profile of health and safety planning and management throughout the construction project.

## Health and safety management system

The most appropriate way for the principal contractor to address the requirements of construction health and safety is to establish within its organization a health and safety management system. This might be a dedicated system or one encompassed within other organizational systems, such as total quality management. By adopting this approach the organization is likely to generate the necessary company policies, culture, procedures and practices to ensure that health and safety is given the attention and respect that it truly demands. Moreover, effective health and safety management will be based on a sound corporate system, with project procedures which consider health and safety as a major contributor to the organization's holistic success.

A concerted approach to health and safety management in construction is absolutely essential. Government has recognized and the industry accepted that the undesirable accident record of construction must be improved. The CDM Regulations place clear and onerous responsibilities upon the main contractual parties to deliver effective health and safety management. It is suggested throughout Part C of this book that the implementation of a clearly conceived, well-structured and highly organized health and safety management system is the most appropriate way for any principal contractor to ensure that it makes a full contribution to providing a safe construction process.

## Key points

This chapter has identified that:

- The CDM Regulations (HSE, 1994) focus on the management of health and safety throughout the total construction process.
- The CDM Regulations require the establishment of a health and safety plan, delivered in two parts – (1) the pre-tender health and safety plan, and (2) the construction phase health and safety plan.
- A well-formulated project health and safety management approach will provide for the purposeful interaction between the two-part process.
- An appropriate way for the principal contractor to address the requirements of health and safety management and meet its obligations under the CDM Regulations is to establish a health and safety management system within its organization.

## Reference

Health and Safety Executive (HSE) (1994) *The Construction (Design and Management) Regulations 1994*, HMSO, London.

# 11 Organizational framework

## Introduction

This chapter focuses upon the consideration which needs to be given to establishing an effective organizational framework to underpin successful health and safety management. Contracting organizations, like organizations in many other business sectors, have had to respond to considerable changes within their respective marketplace. In so doing, many have established management systems in response to particular demands, such as quality assurance, environmental impact and, indeed, health and safety. It is not uncommon for the term 'system' to be confused with 'management'. In this chapter the concept of the organization establishing a parent management system and specific management concepts being the subsystem functions, of which health and safety is one important function, is presented. The chapter provides the basis for appreciating the fundamental aspects and issues associated with organizational framework and sets the scene for considering management structure and health and safety management system development.

## The influence of change

There has been considerable change in the orientation and practice of management by organizations in all sectors of business over the last 20 years. The customer focused marketplace and highly competitive business positioning has demanded clear attention to performance improvement and added value delivery. Changing and more demanding business environments have highlighted the need for organizations to have dynamic and strong strategic leadership coupled with robust and effective directive and operational management. Organizational change appears to have become synonymous with outsourcing, downsizing and re-engineering. Such practices have been widespread and their effects have often been radical, severe and not without some degree of resistance within many organizations.

There has been a considerable culture shift from the morphostatic, or inward looking, organization, with its traditional and often outdated business practices, to the morphogenic, or open-minded, organization in which change is proactively championed and even revered as being prerequisite to business dynamicism (Griffith and Watson, 1999). Such change has influenced most significantly employment structures and the skills and attributes valued by organizations (Anderson and Marshall, 1996). In the 1970s, bureaucratic type structures and strictly prescribed job specifications valued in employees the educational basics such as literacy, honesty and reliability. In the 1980s, a greater focus on business operations and in particular the drive for quality led to a preoccupation with personal skills development such as communication, self-desire and assertiveness. In the 1990s, the flatter lean-organization culture, where more employees were responsible directly for parts of the business, led to a greater need for commercial and entrepreneurial vision and the targeted delivery of outcomes by empowered project managed teams.

With the evolution of organizations into the morphogenic type, external influences become equally as important as intra-organizational aspects to structure and human resourcing. Greater cohesiveness in management is needed to accommodate more demanding customer requirements within the marketplace. Competitiveness must be better understood to ensure the added value of products or services. Increasingly stringent legislation, for example in environmental matters and in particular health and safety, has had to be recognized and responded to. Such important factors require that organizations are clear in the designation of corporate management roles and that project managers are supported in their enabling and empowering functions.

Human resource management has evolved from traditional personnel management. It advocates a holistic approach, where personnel matters are given greater attention by line managers and are considered with direct reference to the core business planning of the organization. Emphasis is given to the business objectives and relating these to the performance management of teams and individuals. Bringing the efforts of the different business team strands together forms the basis of organizational systems management.

## Construction organizations

Within many organizations the structure and organization of a management system and its parts, or subsystems, are relatively uncomplicated. This is because many businesses are based on a single and central corporate management location with single, or a small number of, production sites. The process of construction is not wholly unique as its constituents are repeated from project to project. However, construction organizations differ from other organizations in well recognized and accepted ways, for example in their geographically dispersed decentralized management, temporary organizations, multi-project situation, and interdisciplinary elements. Such factors often dictate that different, although interrelated, management systems must be established for the corporate management organization and for the management of each project organization. In effect, two levels of management are established. Within a contracting organization these levels are manifest in a parent management system established at corporate level which is translated into procedures and working instructions at project level.

The effective management of both the corporate organization and the various project organizations it supports is dependent upon the synthesis of many interdisciplinary elements and resources. It is in the functional interaction of all the parts that produce the necessary synergistic effects that lead to the successful delivery of the construction project and hence contribute to the well-being of the corporate organization.

## Management systems

Systems management concepts have permeated management thinking for well over two centuries. To some extent a systems approach can be advocated in any organization to meet many management requirements as it meets a number of the most basic organizational needs for developing structure and approach.

A management system is, simply put, 'a way of doing things'. A system develops protocols and procedures which bring structure, order, and therefore stability, to an organization where otherwise there might be a potential for chaos. If configured as a holistic and morphogenic open system, i.e. one which interacts with its surroundings, it can meet wider organizational needs. In practical terms, a system represents a hierarchy of guidance to management presented in a manual, sets of procedures and working instructions, which guide the organization in applying management concepts, principles and practices.

To meet the basic needs for organizational existence and development any management system must itself meet several fundamental requirements. A system should in essence be simple but not be simplistic, i.e. it should be easy to understand, interpret and implement by the people who work within it and interact with it. It should give reliable and consistent outcomes, and must be capable of being translated into easily implemented procedures and working instructions. The latter is particularly important within construction as at corporate management level the system is function orientated and at project site level it is task based.

## System and subsystem relationship

Within organizations the term 'system' can be misinterpreted, even misunderstood. It is not uncommon for 'management' to be confused with a 'system'.

The system is really the organization that exists to serve the core business. It is represented by a 'parent' management system within which specialist management concepts are addressed. Management is the control of the subsystems, which are usually structured into functions around the specific management concepts, for example health and safety management, quality management or environmental management. Specific concepts such as these are brought together through the parent management system to contribute towards and support the core business of the organization. It is therefore, more appropriate to denote the corporate system as the parent system and health and safety management as a subsystem.

Figure 11.1 illustrates the interrelationship of various organizational subsystems to an organization's parent system, highlighting the application of health and safety management. It can be seen that the core business of the organization is affected by a great many internal and external influences. These influences determine the main parameters for the existence and operation of the organization. The core business can be serviced and supported by as many subsystems as are needed. Each subsystem is essentially a specialized and dedicated management function, for example, environmental management, purchasing, marketing, or quality management. These are vital strands to both the management of the corporate organization and the construction project based organization. All subsystems are governed by the organizational structural and operating parameters set by the internal and external influences. A subsystem will have management elements common to other subsystems and will also have elements unique to itself which result from legislative or marketplace influences. For example, all subsystems should be guided by a clear statement of organizational policy while health and safety management would, in addition, have dedicated management elements meeting the particular requirements of the CDM Regulations (HSE, 1994), for example, risk assessment.

The management elements in each subsystem provide structure and organization to the human resources which support that subsystem. To enact these elements sets of procedures and working instructions will be established at both the corporate level and project level. For this reason, if no other, management systems can best be described as organizational protocols and procedures which personnel implement in accomplishing their roles.

Within many organizations, and within contracting organizations in particular, two distinct levels of management within each subsystem will be developed. One will deliver specific management functions throughout the corporate organization while the second will deliver specific functions at the construction project level. Essentially, the macro subsystem is established at the organization's head office and within this framework a micro subsystem is established for each construction project site. A project plan will be developed for each individual construction project. The interrelationship between the macro and micro subsystems is significant. Systems and subsystems must be

# Construction health and safety management

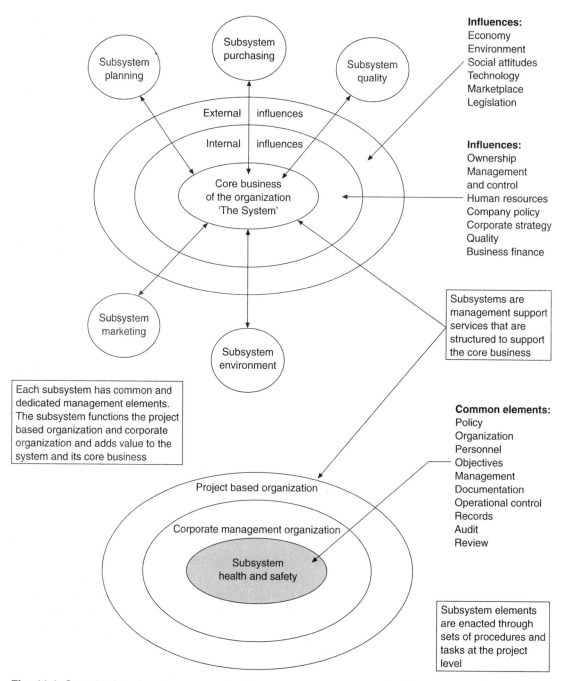

**Fig. 11.1** Organization viewed as subsystem 'management support services' supporting the system and its core business

interactive in the widest sense. There must be close compatibility between the two as each is mutually supportive and must link in with and support the core business. In addition, there must be commonality and consistency in approach within the various subsystems as each should have an ability to share information across the boundaries with other subsystems. This might involve, for example, passing information from one project site to another, or the environmental subsystem sharing information with the quality subsystem.

## Influence of management systems on organizational framework

Establishing a parent management system and focused subsystems within a company has a fundamental influence on its structure and organization. As a system apportions responsibility throughout the entire organization from corporate level to project level the most obvious influence of a system is upon the organization of management. Figure 11.2 illustrates the organization for management of any subsystem.

In Figure 11.2 management is clearly stratified into three broad levels: strategic, directive, and operational.

The role of strategic management is vested in executive management whose primary function is to sanction subsystem development and provide organizational policy and leadership and guide directive and operational management towards supporting the main system, that is, the core business of the organization.

Directive management is charged with developing and implementing the subsystem elements. This will include, for example, the organization of personnel, administering documentation and sets of procedures, training for staff, and auditing the implemented procedures.

Operational management focuses upon the project site and the day-to-day activities of supervisors who implement procedures and working instructions and who monitor and report on performance.

Organization of this kind is typical of subsystems structure within construction industry contracting organizations. This can be seen in practice by many companies in the structure and organization of, for example, quality management or environmental management. The organization and management of the construction process lends itself to a systems approach through specialization.

In recent times traditional specialists have been joined by a complement of managers serving more recently accepted concepts such as quality assurance, environmental impact and health and safety. Construction could not operate without such specialization. Each specialist contributes through implementing particular sets of procedures reinforced by long-established professional working practices. Specialists guide the various stages of the total construction

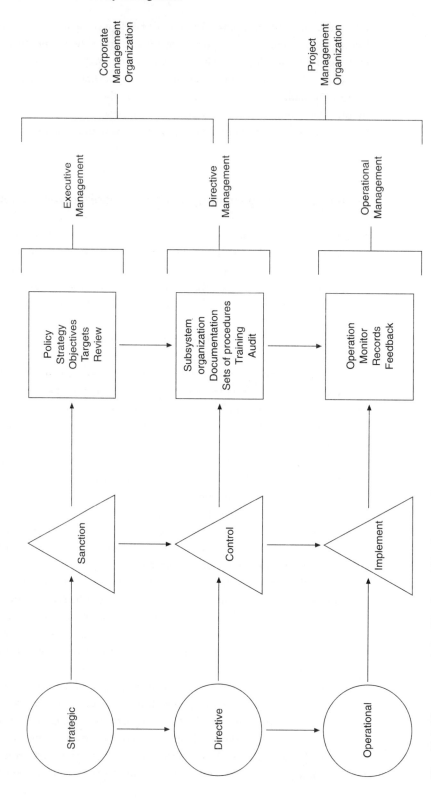

**Fig. 11.2** Management stratification within the organization of a subsystem
*Source*: adapted from Griffith (1994)

process, and the management functions, or subsystems, that they implement help them to execute the many, varied and complex arrangements that are needed to make the construction process work in practice. Furthermore, today's construction clients and the corporate management of construction organizations require an assurance that the company and project procedures are being followed. This is particularly important where independent monitoring or auditing is used to determine organizational performance (Griffith, 1997).

## Subsystems as management support services

The many subsystems used within the construction processes must be seen in context. Each represents a management support service, that is, a service which can be paramount to project team effectiveness, but, which is not in itself a producer of direct financial income. Support services augment both the corporate organization and each temporary project organization by providing a service. They interact with, influence and are prerequisite to maintaining the core business by linking in to the company's parent system. To be cost effective to an organization each subsystem must be efficient and effective in serving both the project organization and the corporate organization.

It is unfortunate that the same specialization that can contribute so effectively to project success can so easily create difficulties, typically at the boundaries between specialists. As additional subsystems are introduced to the construction process to meet more demanding requirements, the boundaries between the subsystems can be indistinct resulting in procedures which can become muddled. When this happens the intended benefits of the management subsystem can be lost, affecting both the cost-effectiveness of the individual project and the well-being of the corporate organization. As poor performance becomes a reality on one construction project so pressure is placed on other projects for greater success in compensation. From a corporate perspective such failure invariably means that organizational overheads must be redistributed across a narrower spread of business, so reducing profit margins.

The best way to ensure that management subsystems are both efficient and effective is to ensure that all personnel understand clearly the role that the subsystems are fulfilling. The subsystems exist to structure and organize management and working procedures at both corporate and project levels and these procedures exist to serve the organization's core business.

Such an approach requires an astute ethos and culture to be developed within the organization. A systems philosophy requires the organization to:

- Adopt a holistic, or whole, organization culture in which management drive and commitment comes from the top and is actively encouraged to permeate the entire organization.
- Present a well-defined framework within which to organize and deploy effectively the resources necessary to support the management subsystems.
- Formally structure its management system and subsystems to develop uniform procedures which support activities across the various levels of management.
- Be proactive in developing the necessary dynamic characteristics to ensure that subsystems interact effectively with others and deliver requisite service at both project and corporate levels.

## Influence of management systems on human resources

Many companies within the construction industry have, in recent years, tended to rationalize and to simplify their organizational structures. Management activities which have proved to be less than vital to the core business have been restructured or removed and many nonessential functions have been shed with the resulting loss of jobs. Contracting organizations, for example, are competing fiercely on costs with little if any margin for profit-making, are niche marketing their services more than ever before and must respond more flexibly to increasingly demanding client requirements. There has also been a tendency towards subdividing the business into independently managed strategic business operations founded often on functional specialization where management have fully devolved responsibility for business performance. They are accountable to but are supported less by the corporate organization.

Many traditional, contracting organizations have become more adept in subcontracting. This effectively reduces their direct responsibility for people-management within the project organization. It does, of course, not eliminate the management function as their subcontractors must be effectively co-ordinated and managed.

Certainly the changing nature of organizations that work within the construction processes has promoted leaner organizations with a greater degree of specialization. These characteristics have tended to lend themselves to a systems management approach to organization and operation. With this, there has been a refocus upon the ways in which human resources are regarded and valued by the more morphogenic organizations within construction.

Management and workforce groups tend to be organized into informed and creative critical mass elements which the organization regard as assets. Managers of such groups are, more than ever before, empowered to take responsibility to organize, motivate and lead these resources in managerially devolved activities. Effectively they run their own business units within overall business parameters set by the parent organization. Such a focus of human resource activity has led to stronger corporate and project foci for implementing structured support services for employees, for example 'Investors in People' (IiP). They also have a clearer direction for business activity to be more performance led and output measured. The added value dimension of delivering customer satisfaction is uppermost and both the organization of the business and the activities of the human resources is directed to this purpose (Griffith, Stephenson and Watson, 2000).

Management systems demand that organizations focus on specific activities to underpin the contribution of human resources. Management subsystems can only be efficient and effective where, for example, clear communication, responsibility, and feedback mechanisms are in place. In addition, because systems often introduce new procedures, working practices may need to be redefined and appropriate training given to both management and employees. This is why training is always included in any management system development. Organizational commitment to training can alleviate many of the problems associated with handling change. Personnel may fear new procedures and the perceived intent behind them and appraising and training managers and operatives will help clarify misconceptions, allay potential fears and avoid organizational dysfunctionality. Where a management system is successfully introduced, the real investment by the organization is not so much in developing the policies, organization and procedures but rather in developing the human resources that implement and support the system.

## Commitment to management systems

A classic dilemma of a systems approach is that as attention is given to a particular subsystem, or specific elements within a subsystem, the integrity and efficiency of the holistic organizational system is often compromised. In such circumstances many of the intended benefits of the individual subsystems may be lost as each subsystem competes for appropriate attention, adequate resources and management commitment.

A prerequisite to developing any approach is that it should achieve a gain in synergy to the corporate organization in pursuit of the core business, that is, the whole must be greater than merely the sum of the parts. The various management subsystems must seek to establish synergistic linkages with other subsystems and the organization's parent system. At the same time the subsystems must also recognize their own distinct and often necessary differences. These differences will, for example, respect particular legislation or the formal system specifications of industry sector accreditation bodies. For example, a health and safety management system must meet the requirements of the CDM Regulations (HSE, 1994) and, ideally, BS 8800 (BSI, 1998). An environmental management system should ideally meet the requirements of ISO 14001 standards (ISO, 1996).

Unfortunately, achieving optimum synergistic benefit is far from straightforward. A number of key reasons predominate. First, the tradition within construction for specialization means that the structural separation of management functions can be counterproductive to effective resource allocation. Second, the divide between the two levels of management within a subsystem, i.e. the corporate level and the project level, can make subsystem interaction less than effective. Third, where there are a number of independent management subsystems in operation there will always be some level of competitive demand for resources or priority attention. Given these characteristics, which frequently occur within construction, synergy can clearly be difficult to achieve.

## Key points

This chapter has identified that:

- Two discernible yet interrelated levels of management need to established within a contracting organization to provide a basis for health and safety, these are:
  - a parent management approach at corporate level
  - procedures and working instructions at project level.
- There are good reasons to develop a formal health and safety management system within an organization in meeting the two levels of management.
- It may be useful and purposeful to view a system as the organization with management functions, of which health and safety is one, forming subsystems that serve both the project organization and the corporate organization.
- Establishing management functions, or subsystems, has a fundamental influence upon structure, organization and human resources with three distinct levels being important to successful system implementation: (i) strategic, (ii) directive, and (iii) operational.
- A health and safety management subsystem exhibits elements in common with other management functions and also has particular requirements determined by legislation and the professional practices required.

## References

Anderson, A. and Marshall, V. (1996) *Core Versus Occupation Specific Skills*, Department for Education and Employment (DfEE), HMSO, London.

British Standards Institution (BSI) (1998) *BS8800: Specification for Health and Safety Management Systems*, BSI, London.

Griffith, A. (1994) *Environmental Management in Construction*, Macmillan, Basingstoke.

Griffith, A. (1997) 'Towards an integrated system for managing project quality, safety and environmental impact', *Australian Institute of Building (AIB) Papers*, Vol. 8, pp. 67–77, Melbourne.

Griffith, A. and Watson, P. (1999) 'Optimising management systems for construction', *Australian Institute of Building (AIB) Papers*, Vol. 9, pp. 39–49, Melbourne.

Griffith, A., Stephenson, P. and Watson, P. (2000) *Management Systems for Construction*, Longman/CIOB, Harlow.

Health and Safety Executive (HSE) (1994) *The Construction (Design and Management) Regulations 1994*, HMSO, London.

International Standards Organization (ISO) (1996) *ISO 14001: Environmental Management Systems – Specification with Guidance for Use*, ISO, Geneva.

# 12 Management structure and management system

## Introduction

This chapter focuses upon the consideration which needs to be given to establishing an appropriate management structure and management subsystem for effective health and safety management. Unlike many manufacturing, product, or services sector organizations, contracting companies need to structure their organization and activities at two levels. They must structure the corporate organization, usually through the development of a parent management system, and also the construction project organization, of which there can be many and which can be greatly geographically dispersed and exhibit many individual features. This chapter provides the basis for appreciating the fundamental aspects and issues associated with management structure and systems and provides the foundation for considering the development of a health and safety management subsystem.

## Management structure

Figure 12.1 illustrates a typical management structure within many medium to large contracting organizations. As explained in Chapter 11, the holistic organization structure is divided into two levels. These are the corporate organization structure and the project organization structure.

The management structure indicates that the ultimate responsibility for the company's performance in all areas rests with the Board of Directors, headed by the Managing Director. Below Board level, the organization is served by a collective of sections or departments, for example, finance and legal, technical services, marketing and public relations, commercial services and project (production) management. These are augmented by company support services, which include personnel, education and training, and management support services, which include health and safety management.

Such companies are often divided geographically and into divisions, for example, housebuilding, building projects and civil engineering works, each headed by divisional and regional directors. Sections or departments are each managed by a head who provides management support to both the corporate and project organizations.

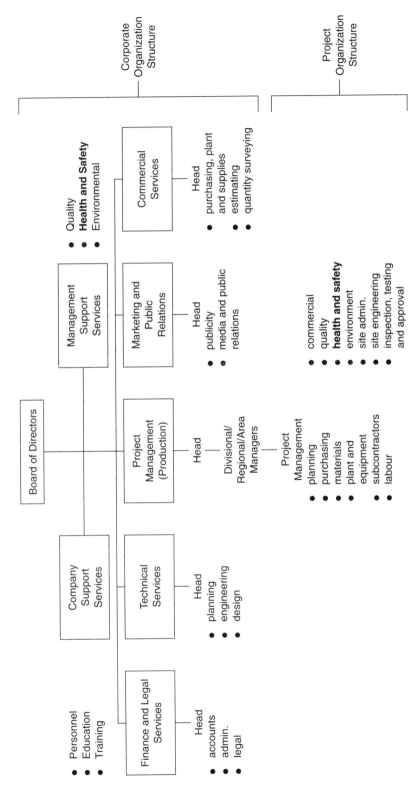

**Fig. 12.1** Management structure for corporate and project organization within a constructing company

At project organization level each project will be headed by a construction project manager who draws support from each corporate service and the management services based on site. These are shown in Figure 12.1 with health and safety management services highlighted. A health and safety system manager, where structured as such, will be responsible for the implementation of the management system procedures and the documentation and records relating to those procedures. In addition, the manager will be responsible for evaluating the data collected and informing senior management of any deficiencies identified in the operation of the management system.

Where a dedicated health and safety management system is not used and health and safety falls within the remit of another organizational system, for example quality, then a joint support services manager would be responsible for overseeing health and safety within the management of the quality system.

Utilizing the process of auditing (see Chapter 19), the Board of Directors, in conjunction with the management support services head of health and safety, or joint support services head, maintains a system of monitoring, review and system development and update. This is essential to maintaining management systems which purport to be dynamic and improving.

## Management system

**The parent system**　A parent management system is structured to provide an operational framework for the management of the whole organization. Its focus is to support the undertaking of its core business, the assurance of its quality, safety and contractual obligations and give it the optimum control of its business activities.

The system will ensure that key objectives are met as follows:

- Specified requirements and standards are developed at key stages throughout procurement, design, construction and maintenance of a project.
- Materials, services and all construction activities comply with the specified requirements and standards.
- A mechanism of review and improvement is applied to all stages of the construction process.

## Link between management structure and the system

To achieve the above key objectives the organization requires the establishment of a management structure, as outlined on pp. 116–18. Using this structure the company can develop a holistic management system within which the required management support services can be established. As discussed in Chapter 11, a dedicated health and safety management subsystem can be developed or health and safety could form an integral part of a quality management subsystem.

## Documented (formal) system

For any management system or subsystem to be effective it must be formalized in some way, that is, it should be in a structured and documented form. The usual way in which this is achieved is for the company to develop a management systems manual document.

Documentation for any system should be structured in such a way that it embraces the activities of both the corporate organization and its project organizations. In this way the system will be consistent with the management structure described previously and illustrated in Figure 12.1.

Almost all management systems are characterized by a collection of documentation which describes system implementation at various levels within the organization. Typically there can be four levels of documentation.

1. The management system manual.
2. Sets of management procedures.
3. Working instructions.
4. Project plans.

As shown in Figure 12.2 the level of detail addressed by the system generally increases from the management system manual to the procedures, to the working instructions. These are augmented by project plans which address the specific requirements of each project undertaken. These project plans will be made up of subsidiary plans to address each management support service area, for example, a health and safety plan, an environmental plan, a quality plan, etc.

### The system manual

The system manual defines the policy, management responsibilities and organization structure which is to be utilized by the company to ensure its ability to meet the specified requirements of its business. It identifies and outlines the skeleton of procedures by which the organization will operate. It presents specific procedures for particular activities, for example, health and safety management or the projects it undertakes.

### Management procedures

The management procedures outline the actions to be followed in key areas of organizational activity, for example, support service management (within which

Level 1: MANAGEMENT SYSTEM MANUAL

- Policy
- Organization
- Responsibilities
- Management Procedure References

Level 2: MANAGEMENT PROCEDURES

Support Service Management
- Quality
- **HEALTH AND SAFETY**
- Environmental

Company Support Services
- Personnel
- Education
- Training

Project Management
- Client/Design Liaison
- Contract Establishment
- Project Management
- Post-contract Review

Service Management
- Finance and Legal
- Technical Services
- Marketing and Public Relations
- Commercial Services

System Management
- Document Control
- Auditing
- System Review
- Feedback and Improvement

Level 3: WORKING INSTRUCTIONS
(examples)
- **Health and Safety Assessment and Control**
- Environmental Assessment and Control
- Quality Assessment and Control
- Plant and Material Procurement
- Subcontract Procurement

Level 4: PROJECT PLANS
(example subsidiary plans)
- **Health and Safety Plan**
- Environmental Plan
- Quality Plan
- Temporary Works Plan
- Project Plan (Programme)

**Fig. 12.2** Various levels of documentation for management systems development

health and safety management is structured), company support services, service management, system management and project management (see Figure 12.2).

### Working instructions

The working instructions define the specific actions to be undertaken when carrying out given operations within the management procedures (see Figure 12.2). Working instructions can be generic in that they are applicable in any situation of carrying out out an operation. They can also be influenced by particular situations, circumstances or environments, in which case additional information will be provided in the project plan.

### Project plans

These are plans which define and describe the characteristics of the particular project. These are the pre-tender health and safety plan and the construction health and safety plan. Details are given in Chapter 17.

## Health and safety management subsystem

**Key elements**  It can be said that the features of effective health and safety management are similar, if not identical, to what might be termed simply as good management practice. The general principles of such practice form a useful basis for structuring a health and safety management subsystem.

The key elements in almost all management systems are:

- Policy.
- Organization.
- Planning.
- System implementation (including monitoring performance).
- Auditing (including review and feedback).

These key elements, shown in Figure 12.3, form the outline structure of the health and safety management system. The same elements can also be employed in the development of other management support services. In fact, they always form an integral part, for example, of quality management and environmental management.

A further key element is essential to health and safety management. This is risk assessment (see Figure 12.3). It is key because it is a specified requirement of construction health and safety legislation under the CDM Regulations, 1994 (HSE, 1994).

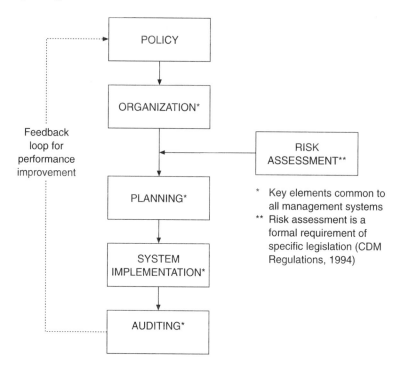

**Fig. 12.3** Key elements of a management subsystem

### Policy

Policies on health and safety influence almost all organizational activities and decisions. They have a bearing on the selection of resources, organization and operation of working, delivering of products and services, and moreover they develop the supporting ethos and culture within which all of these are achieved. Policies for health and safety drive the desire for a safe working environment in which both human and physical resources are safeguarded. This helps to maximize cost-effectiveness in safety terms, which is an important consideration to developing a total loss approach. Policies serve, most importantly, in meeting an organization's responsibility towards current legislation and public accountability.

### Organization

Organization has a considerable influence upon developing a sound and positive culture which allows safety management to be effective. Understanding of the roles and responsibilities of all parties involved is vital to effective health and safety control. Within the principal contractor's organization at both corporate and project levels there is a need for proactive management and leadership to translate procedures and working instructions into effective practices.

### Risk assessment

Under the CDM Regulations all parties involved with the construction processes have a duty and responsibility to identify possible hazards and risks, take appropriate mitigating action and ensure a safe working environment for all persons employed on the project. A structured and formal approach to risk assessment is therefore a key element of a health and safety management system.

### Planning

Effective planning is prerequisite to effective health and safety management. The CDM Regulations require that the client develops the pre-tender health and safety plan to aid the consideration of health and safety during the tendering process, while the construction health and safety plan developed by the principal contractor identifies risk and considers control measures for the construction phase. Planning throughout the development and undertaking of a construction project is, therefore, a key element in health and safety management system establishment.

### Implementation

The most important element of a health and safety management system is implementation. This is quite simply because implementation translates the management system – the manual, procedures, working instructions and project plans – into applied practices. Without effective implementation any management system is totally ineffective. System implementation is specific to the nature of the business and its constituent parts. Within construction, system implementation focuses on the production processes and those key elements which ensure that safe working practices on site are developed and maintained.

### Auditing

Learning from the experience of system application is an important element in health and safety management. Gathering, reporting and reviewing information on project performance in a systematic and reliable way is essential. The most appropriate way to ensure that this takes place is to utilize an auditing mechanism which assesses health and safety performance against predetermined indicators. Auditing is therefore a key element of system development.

**Subsystem framework**   Figure 12.4 illustrates the developmental framework for the health and safety management subsystem. The relationship of the hierarchy of documentation to the key elements of health and safety management can be seen as the input from the corporate dimension to the project level management dimension. The corporate elements include the management system manual and sets of procedures which prescribe the approach to be taken in managing policy,

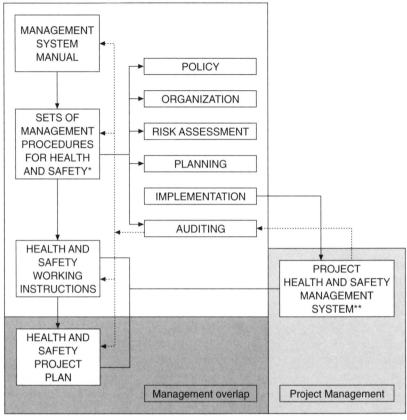

**Fig. 12.4**
Developmental
framework for a
principal contractor's
health and safety
management
subsystem

\*    Other management support services, e.g. quality, environmental
······ Feedback routes for system improvement and development
\*\*    See Chapter 18: System Implementation

organization, risk, planning, implementation and auditing. Working instructions guide generic practices to be adopted while undertaking the site works,
while the project plan guides aspects specifically applicable to the individual
project. Chapters 14–19 describe the important aspects of each of the management system elements that form the subsystem set out in this chapter.

**An outline for
subsystem
development**

It is not within the scope of this book to provide a detailed content, format
and presentation of documentation that an organization might use in developing a health and safety management subsystem manual. Each organization
will have its own way of doing things and therefore must develop its own
systems and subsystems to meet its own particular circumstances. Notwithstanding, it is useful to set out the fundamental skeleton of headings around
which a subsystem can be considered and this is shown in Figure 12.5.

1. **Manual**

- Title page
- Contents
- Preamble
- Instructions on use
- Version
- Circulation list

- Policy statement
  - Policy
  - Standards
  - Ownership and commitment
  - Relationship of manual to procedures and instructions
  - Duties of staff and employees

- Introduction
  - Documentation
  - Manual
  - Management procedures
  - Working instructions
  - Project plans
  - Management
  - Verification
  - Development

- Definition of terms
  - Standards
  - Terms used in system documents

- System documentation
  - Manual
  - Procedures
  - Working instructions
  - Project plans

- Structure and organization
  - Corporate structure
  - Board of Directors
  - Main departments/divisional structure
  - Management support services
  - Company support services
  - Project organization
  - System management

2. **Management procedures**

Management procedures (e.g. for projects)

- Client relationship
  - Responsibilities
  - Commitment
  - Development
  - Maintenance
  - Review

- Contract establishment
  - Responsibilities
  - Tender handover
  - Post-tender review
  - Project plan
  - Planning meeting
  - Programme
  - Site engineering and establishment

**Fig. 12.5** Parent management system and health and safety management subsystem development headings

- cost planning
- project income
- quality
- environment
- health and safety

- ● Project Management
  - project strategy
  - planning
  - purchasing
  - materials
  - plant and equipment
  - subcontractors
  - labour
  - programme
  - cost control
  - quality
  - environment
  - health and safety
  - administration
  - inspection and approvals

- ● Post Contract Review
  - auditing
  - evaluation
  - information for future projects

3. **Working instructions**
   - ● Working Instructions, e.g.
     - health and safety inductions
     - risk assessments
     - environmental control
     - quality control
     - health and safety control
     - toolbox management
     - site instruction
     - site meetings
     - plant use
     - material requisition
     - work inspection and supervision
     - working in hazardous conditions
     - permits to work
     - hot works
     - administration and record keeping

4. **Project plans**
   - ● Health and Safety Plan
     - company safety policy
     - management responsibilities
     - company safety advisor
     - project (contract) manager responsibilities
     - site safety supervision and inspections
     - statutory obligations
     - safety method statements
     - risk assessment
     - first aid
     - accident registers
     - investigation and report of accident incidents
     - induction and training
     - permits to work

**Fig. 12.5** (*cont'd*)

## Prerequisites to management subsystem development

Synergy within any subsystem can only have a chance of success if the utmost commitment is given by the organization. There are a number of prerequisites to management systems development. These apply at corporate level and at project level and focus on sound management, commitment, policy making, management planning and organization.

To ensure that any subsystem has an optimum chance of being effective a number of actions must be made explicit. These actions apply to both corporate and project levels and are detailed below.

**Corporate level**   At corporate level prerequisites to effective subsystem development include the following:

- Demonstrable commitment to organizational policy and strategy.
- A clear statement of organizational ethos and policy, which should be circulated throughout the organization.
- Employee ownership through involvement in formulation, development and implementation.
- Identified goals and targets, at both corporate and project levels, against which performance can be reviewed.
- Adequate resources to facilitate the robust development of framework, structure and organization.
- Appropriate and continuing education and training for management and the workforce.
- On-going review and improvement to applications to enhance experience and expertise.

**Project level**   At project level prerequisites to effective subsystem development include the following:

- Identification of key issues that need to be addressed on the project site.
- A clear mechanism to undertake effective risk assessment.
- Development of action plans in response to identified needs.
- Distribution of good practice guidelines to all staff.
- Determination of audit procedures between the corporate system and the project subsystem.
- Briefing and training of all project team members on procedures to be implemented.
- Working instructions detailing the procedures to be adopted at operational level.
- Item checklists for the monitoring of specific aspects.
- Self-audit and review documents for subsystem supervisors.

- Guidance notes on potential actions when problems occur.
- References to corporate management where assistance may be needed by the project organization.

It is absolutely essential that the organization carefully considers its subsystems and the deployment of human resources. Both management and workforce need a subsystem that is efficient, effective, and clear and easy to understand and implement, if benefits are to be achieved, organizational synergy to be maximized and the core business to be supported.

## Key points

This chapter has identified that:

- Within most medium to large principal contracting organizations management is divided into two structured levels: (1) the corporate organization, and (2) the project organization.
- The organization is managed by a collection of services structured into functional departments augmented by company support services and management support services.
- Many organizations utilize a 'parent' management system which is structured to provide the operational framework for the management of the whole organization.
- Management support services, structured into specialist subsystems, underpin the parent system. Health and safety management is one such subsystem.
- Systems and subsystems are practice instructions for management formalized in a collection of documentation which includes manuals, management procedures, working instructions and project plans.
- Within any system there are key elements of activities which must be carried out. These are policy, organization, planning, implementation, and auditing. Risk assessment forms a further key element as it is specified under government legislation for health and safety management.
- While any organization should develop its own management approach to corporate and project organization and health and safety management, a skeleton of outline headings for system development can be identified.

## Reference

Health and Safety Executive (HSE) (1994) *The Construction (Design and Management) Regulations 1994*, HMSO, London.

# 13 Human aspects

## Introduction

A successful health and safety management approach will recognize and value the important and fundamental contribution that individuals make in support of a safe working environment. Effective health and safety policies seek not only to combat the possibility of accident and hazardous incidents, but also to generate a wider culture of welfare, health and safety throughout the organization. To achieve this the organization must seek to understand the key influences upon the behavioural characteristics of employees and management. The development of behaviour safety management approaches has been the subject of study in recent years leading to goal setting practices and the piloting of behavioural safety auditing. This chapter discusses the importance of people to an organization developing a health and safety management subsystem.

## The importance of people to the organization

Crucial influences upon the establishment of effective health and safety management practices are the perceptions of health and safety aspects held by employees and the attitudes projected by the employer. A health and safety management approach will be best demonstrated where people generally, and the individual specifically, are regarded as the focus of safeguards and are involved in providing and maintaining those safeguards. Good health and safety practices can perpetuate not only good ongoing practices, but also enhance work enjoyment, individual and group satisfaction and lead to improved productivity. People who are committed and interested will undoubtedly work in an environment of greater awareness and take care not only for themselves but for others.

Successful health and safety management subsystems will recognize the important contribution that individuals and groups can bring to developing an aware and safe working environment and build their policies, organization and practices around this. The most effective health and safety management policy will seek not only to prevent accidents and consequent injury to

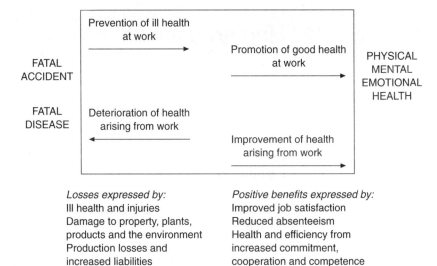

**Fig. 13.1** Spectrum of occupational health *Source*: adapted from HSE (1994)

*Losses expressed by:*
Ill health and injuries
Damage to property, plants,
products and the environment
Production losses and
increased liabilities

*Positive benefits expressed by:*
Improved job satisfaction
Reduced absenteeism
Health and efficiency from
increased commitment,
cooperation and competence

persons but to generate a wider culture of health and welfare within the corporate organization and its various multiple project organizations. Promoting a positive approach is based upon fostering the belief that people, both individually and collectively, are an essential resource and contributor to the core business of the organization.

## Total loss approach

The ultimate goal of an organization employing a health and safety management subsystem is to create a corporate culture and project environment in which welfare standards are high and accidents and hazardous incidents are minimized. This represents the human dimension to the total loss approach to health and safety management discussed in Chapter 4. The most effective health and safety management subsystem will successfully manage those situations that give rise to human injury, damage to property and production site losses (see Figure 13.1). Moreover, an effective approach will allow the organization to learn from construction accidents and seek improved management control.

## Influences upon employee attitudes

The principal influences upon employee attitudes to construction health and safety come from four directions: (1) the personal human influences of the

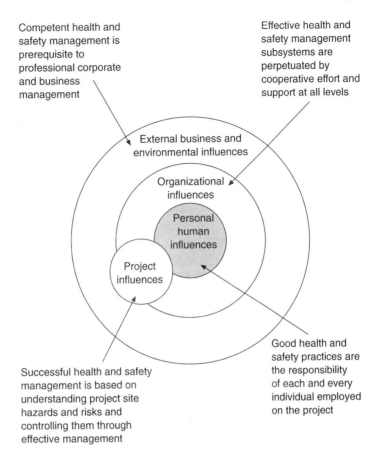

Competent health and
safety management is
prerequisite to
professional corporate
and business
management

Effective health and
safety management
subsystems are
perpetuated by
cooperative effort and
support at all levels

External business and
environmental influences

Organizational
influences

Personal
human
influences

Project
influences

**Fig. 13.2** Influences
upon employee
attitudes to
construction health
and safety

Successful health and safety
management is based on
understanding project site
hazards and risks and
controlling them through
effective management

Good health and
safety practices are
the responsibility
of each and every
individual employed
on the project

individual; (2) the influences of the project; (3) organizational influences
from corporate management; and (4) influences of the environment and the
external business dimension. These are shown in Figure 13.2.

The influences on employee attitudes act both singularly and in combination to present management with multifaceted demands upon both the
corporate and project organizations. Any or all of these aspects can combine
to become contributing influences towards accidents (see Figure 13.3). Understanding the multiple and, at times, complex interrelationships between these
influential dimensions is essential to comprehending the behavioural aspects
of health and safety management. The behavioural characteristics of both the
workforce and management are crucial to establishing an effective health
and safety management subsystem within the organization.

## Behavioural aspects

It is clear that the individual can play a significant role in health and safety
management as a direct result of their awareness, knowledge and behaviour.

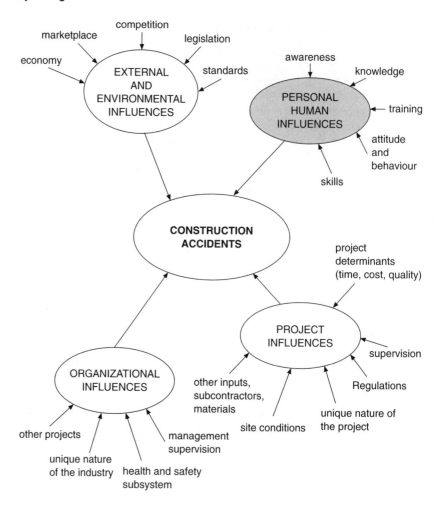

**Fig. 13.3** Some key contributory influences on construction accidents

It is also recognized that often the association of the individual with a health and safety incident or accident may only be a symptom of a much deeper rooted cause. Frequently the fundamental cause lies in the safety policies pursued, the organization structure, the effectiveness of the risk assessment carried out and the working procedures and levels of supervision implemented. This can be influenced by any or all of the influential dimensions described earlier: the external business marketplace; the environment; the organization; the project; and the individuals themselves. Fundamentally, management behaviour is equally as important as the behaviour of the individual. Within a structured health and safety management approach at both corporate and project organizational levels there must be commitment towards developing a culture and subsystem in which the behaviour of management, supervisors, groups and individuals focuses upon critical safety management practices.

## Behavioural safety management

Behavioural safety management is concerned with 'a range of techniques which seek to improve safety performance by setting goals, measuring performance and providing feedback' (Lingard and Rowlinson, 1994). Indeed, all those sound aspects demonstrated by a well-structured and well-organized health and safety management subsystem. Various recent studies of health and safety performance (Lingard and Rowlinson, 1994; Duff *et al.*, 1994; Tang *et al.*, 1997; and Tam and Fong, 1998) have reported on the need for greater awareness of behavioural aspects in health and safety management development. Such work has been of considerable importance both in the UK and particularly in the Far East where accident statistics are much worse than those of Western countries. Local situations are, of course, influenced by particular factors and this is true in the Far East. For example, local industry may utilize a high percentage of labour only subcontractors, or may adopt localized traditional methods of construction and use of materials.

Behavioural safety management focuses upon recognizing that behaviour culture is critical to good safety management, translating this into performance specific practices based on management and operative goals, the assessment of performance, and provision of feedback. Research by Lingard and Rowlinson (1994) has shown that goal related behaviour modification can be highly effective in contributing to improved site safety practices. Particular benefits were seen in improvements to site housekeeping, an aspect of site health and safety well known to be a determinant of incidents and accidents.

## Performance standards

Determining the performance of groups and individuals is essential in translating health and safety policies into effective practices. This involves determining clearly what people need to do, what their responsibilities are, and how their practices will be measured and judged. Behavioural characteristics of employees will be focused on health and safety where the organization, as part of the health and safety management subsystem, gives a clear definition of their roles, duties and responsibilities. This can be effected by:

- Making health and safety an overt part of their job description.
- Implementing a formal performance review which measures and rewards positive commitment and achievement of goals.
- Giving support, help and training where deficiencies are identified.
- Invoking penalties where serious continual deficiencies are apparent.

Such an approach requires the organization to develop and implement a positive and continuous method of assessment and review. One such approach is a behavioural safety auditing procedure.

## Behavioural safety audits

The vexed issue of behaviour modification approach is how to implement goal setting effectively as part of a structured health and safety management subsystem. Reports on research by Duff (1994) and Cameron (1998) have pointed the way towards the development of behavioural safety auditing. The purpose of such an approach is to offer a practical tool to ensure that key safety management behaviours are observed, for example safety induction training. Their auditing approach focuses upon systems safety, for instance, safety committees, induction procedures, toolbox talks, inspections, and documented records. The procedures use an audit as a means to develop goals, implement checks and provide ongoing feedback. While this represents good sound management practice the innovative dimension lies in the fact that the audit measures the presence of safety practice and not its absence. The work is a first in that it is 'the first occasion safety culture has been systematically measured in the construction industry' (Cameron, 1998).

The focus is upon a proactive approach rather than a reactive one. A safety culture is sought in which total safety management becomes the overriding goal. This is exactly those characteristics that an effective organizational health and safety subsystem seeks to achieve and the approach advocated throughout this book. It will be seen in Chapter 18 that facets of behavioural safety management are incorporated within an effective health and safety management system.

## Key points

This chapter has identified that:

- A successful health and safety management subsystem will recognize the important contribution that individuals and groups can bring to developing an aware and safe project environment.
- Promoting a positive approach to health and safety management is based upon fostering the belief within the organization that people are a key asset in supporting a safe working environment for everyone.
- The principal influences upon employees' attitudes and behaviour to health and safety management come from four directions: (1) the individual;

(2) the project; (3) the organization; and (4) the environment and external business dimension.

- Personal human influences upon health and safety behaviour include the individual's awareness, knowledge, skills, attitudes, and training and development. These must be channelled into safe working practices on site.
- Management behaviour and operative behaviour are equally as important and should focus upon critical safety management practices.
- The development of performance standards and goals for individuals and groups should be made clear and form an integral part of job descriptions, assessment and review.
- One approach to positive behaviour modification to support good health and safety practice is to consider a behavioural safety audit mechanism (Cameron, 1998) where goals are set, monitored, reviewed and fed back into proactive and ongoing health and safety practices.

## References

Cameron, I. (1998) 'Pilot study proves value of safety audits', *Construction Manager*, 23 April.

Duff, A. R., Robertson, I. T., Phillips, R. A. and Cooper, M. D. (1994) 'Improving safety by the modification of behaviour', *Construction Management and Economics*, Vol. 12 (6), pp. 67–78.

Health and Safety Executive (HSE) (1994) *Successful Health and Safety Management*, HMSO, London.

Lingard, H. and Rowlinson, S. (1994) 'Construction site safety in Hong Kong', *Construction Management and Economics*, Vol. 12 (6), pp. 501–510.

Tam, C. M. and Fong, W. H. (1998), 'Effectiveness of safety management strategies on safety performance in Hong Kong', *Construction Management and Economics*, Vol. 16, pp. 49–55.

Tang, S. L., Lee, H. K. and Wong, K. (1997) 'Safety cost optimisation of building projects in Hong Kong', *Construction Management and Economics*, Vol. 15, pp. 177–186.

# 14 Policy

## Introduction

This chapter focuses on the key factors to consider in developing an effective health and safety policy. Construction organizations which successfully achieve a high standard of health and safety practice invariably have in place clear and accepted health and safety policies. Moreover, these policies embrace the organization's social and corporate responsibilities for health and safety and develop a holistic and supportive culture in which good practice becomes the norm rather than the exception. Clear health and safety policies provide for the systematic and cost-effective identification, monitoring and control of health and safety risk and are prerequisite to the planning and organization of an effective health and safety management system.

## Policy definition

A health and safety policy is:

- a published statement (or set of statements) reflecting the organization's intentions in relation to the management of health and safety matters.

A health and safety policy should:

- Define the organization's corporate philosophy towards health and safety matters, in the context of its business activities.
- Be clearly presented in the form of a policy statement, originating from the organization's board of executive management.

## Policy formulation

Policy formulation is fundamental to the establishment of an organization's health and safety management system. It forms the basis for developing procedures and operational tasks and is influential in the assignment of management responsibilities and the development of human resources. To facilitate management leadership and support and encourage the motivation of employees, the policy must be clearly determined at strategic management level by organizational executives. Thereafter, it must be communicated unambiguously through each level in the organization's management heirarchy. This will be achieved by using understandable sets of procedures and practical working instructions, that is, the implementation of the health and safety management system. The organizational framework document specifying this approach will be the company's health and safety policies and procedures manual. Alternatively, policies and procedures may form a sub-section of another organizational management system.

At corporate level, policies must, in practice, be sufficiently flexible to accommodate changes in organizational circumstances, future developments and evolving legislation. At implementational, or operational, level policies must be sufficiently structured and robust to invoke purpose and discipline, but again allow some flexibility to meet changing circumstances.

From the perspective of directive and operational management, and the workforce, it is crucial that a clear policy emerges. Health and safety management success is often greatly influenced in practice by organizational culture and understanding, employee awareness and attitudes and the desire by project teams and individuals for high standards of health and safety performance.

Although health and safety policy is predominantly the responsibility of senior corporate management, its development must be cognizant of each managerial and workforce level. This is important in allowing the opportunity for the company's corporate philosophy and ethos to permeate throughout the organization. In this way, the policy will become a meaningful and purposeful part of structure and procedures.

Policy should reflect top-down management with commitment coming from, and being seen to come from, executive management. In any management system it is almost always the absence of clearly defined policies that can be traced back to system difficulties and failures.

The scope of policy formulation and the issues that this raises for the organization will depend upon the individual characteristics and constitution of the company and its business. As policy lays the foundation for the development of health and safety practices and performance measurement, the policy should be developed with some thought to measurable criteria as this facilitates goal and target setting. However, it is important at the policy making stage that over-detailing is avoided. Policy should reflect the company's holistic philosophy and corporate commitment to health and safety. It is a pronouncement in the inter-organizational sense and is in the public domain. Fine detail of the translation from policies to practices should be included in intra-organizational documentation such as the health and safety procedures manual.

To be successful, a health and safety policy must be:

- Relevant to the organization's core activities.
- Directed from executive level.
- Translated into practical sets of procedures and working instructions.
- Committed to throughout the entire organization.
- Implemented and maintained in a systematic way.
- Available to regulatory bodies and public scrutiny.
- Amenable to review and improvement.

## Key factors

Arguably, there are many factors that an organization must consider when formulating its health and safety policies. Figure 14.1 highlights eight key factors (listed on the right of the Figure). It is not intended to provide an exclusive list of factors, but to represent *some* important organizational aspects that must be considered.

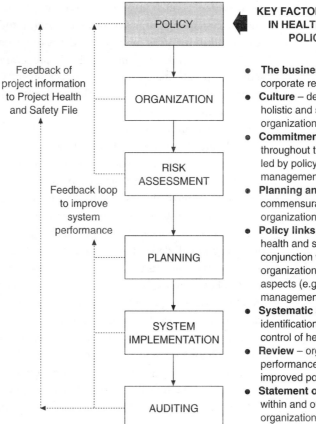

**KEY FACTORS TO CONSIDER IN HEALTH AND SAFETY POLICY MAKING**

- **The business** – social and corporate responsibilities
- **Culture** – development of a holistic and supportive organization culture
- **Commitment** – perpetuation throughout the organization led by policy of executive management
- **Planning and resourcing** – commensurate to meeting organizational policies
- **Policy links** – appreciation of health and safety in conjunction with other key organizational and project aspects (e.g. environmental management)
- **Systematic approach** – identification, monitoring and control of health and safety risk
- **Review** – organizational performance and feedback into improved policy making
- **Statement of policy** – both within and outwith the organization

**Fig. 14.1** Key factors to consider in health and safety policy making

**The business**  The way in which health and safety thinking is incorporated into an organization's decision making processes and business activities is essential to good standards of performance. The foundation for this is sound policy making. The organization has a commitment to consider not only its corporate responsibilities but its social responsibilities. This is especially important within construction since its activities hold considerable potential to present hazards and dangers to the public. In addition to meeting the many statutory obligations and increasingly stringent legislation, an organization has a responsibility for the welfare, health and safety of its employees and also those who interface with, or are associated with, its activities.

There are a host of organizational aspects that would need to be considered during policy making and many impinge upon health and safety policies, procedures and practice. These are shown in Figure 14.2 and may be broadly grouped under the following headings:

**CORPORATE & SOCIAL RESPONSIBILITIES**
- Ethos and philosophy
- Ethics
- Accountability
- Responsiveness
- Organizational effects
- Statement of policy

**BUSINESS STRATEGY**
- Business mission
- Commitment to policies
- Marketplace awareness
- Commercial positioning
- Customer focusing

**HUMAN RESOURCES**
- Staff deployment
- Recruitment
- Redeployment
- Training and development
- Incentives
- Investors in people (IiP)

**MANAGEMENT SYSTEMS**
- Organizational structure
- Implementation systems (sets of procedures)
- Performance indicators
- Resourcing
- Systems development
- Continuous improvement

**COMMUNICATIONS & IT**
- Identification of data requirements
- Use of IT
- Collection and analysis of data
- Communication of information within the organization

**LEGAL & FINANCIAL**
- EC Directives
- International standards
- National legislation
- Investment
- Loss control and risk assessment
- Liabilities
- Insurances
- Planning and budgetary control

**OPERATIONS**
- Quality management
- Selection, design, construction, maintenance of premises (facilities management)
- Safe working procedures
- Promotion of safe environment
- Emergency practices
- Contingency planning

**MARKETING & CORPORATE IMAGE**
- Public image
- Internal image
- Company promotion
- Corporate literature
- Media interfacing and external communications

**Fig. 14.2** Organizational areas for consideration in policy making

- Corporate and social responsibilities.
- Business strategy.
- Human resources.
- Management systems.
- Communciation and information technology.
- Legal and financial.
- Operations.
- Marketing and corporate image.

The most successful health and safety policies will be generated by an organization which considers people as its most important asset. The focus on people, and especially the individual, will provide the ethos around which a positive organizational culture and subsequent commitment can be encouraged. This will drive not only the making of policy itself but the greater awareness for and acknowledgement and acceptance of policy throughout the organization. Health and safety policy making should consider that all employees and the public are susceptible to accidents and are at risk. It should seek to prevent rather than merely react to accidents and positively promote health and safety through the recognition that people are the most valuable resource to the organization.

A natural extension to the focus on people is to consider the organization's additional resources. Invariably, the cost of accidents involves other components such as damage to property, equipment and materials and all the visible and hidden costs associated with them. The HSE recommends a total loss approach (HSE, 1993) where:

> the preservation of human and physical resources is an important means of minimizing costs.

They recognize that:

> the consequences of accidents are often matters of chance over which there can be little control.

Therefore, a total loss approach requires one to learn from the occurrence of accidents and develop more effective controls. This necessitates that the organization develop policies which support a holistic culture and dynamic system of health and safety management.

**Culture**  The development and maintenance of an effective and supportive organizational culture for health and safety management is essential. The policy of making people the focus must be translated into visible support to give currency and commitment. Effective health and safety management will be fostered through the contribution of management and work groups and the individuals within those groups at each organizational level.

One of the most pertinent ways to encourage a strong health and safety culture is to associate the principal objectives of health and safety management with the mission and key goals of the business. This can only be achieved if the health and safety policy has been clearly set and corporate commitment is unequivocal. Moreover, the support given by and practices demonstrated by executive and directive management themselves must be clearly commensurate with the policies that they set.

Health and safety is a holistic consideration. The whole organization will need to share the beliefs and ethos of executive management. It is vital therefore, that health and safety is seen as an aspect which permeates the entire organization and impinges upon all of its activities. Good health and safety practice must become the accepted way of doing things.

**Commitment**

Commitment to health and safety policies and a holistic and supportive organizational culture starts with and must be encouraged by executive management. They must be unswerving in the underlying belief that health and safety management is essential to corporate success. From executive level, managers must share the belief that health and safety culture is based on the appreciation and understanding of hazard and risk to people and that people are the organization's key asset. A commitment to the total loss approach to health and safety will emphasize the prevention of accidents. This will not only reduce human risk but benefit from the cost-effectiveness of accident prevention.

**Planning and resourcing**

It is absolutely essential that the organization resources health and safety aspects appropriately. There is little point in developing policies which cannot be supported by good organizational procedures and practices. The organization must ensure that it has a clear picture of the likely tasks that need to be undertaken to support effective health and safety practices, how the tasks might be undertaken, and who will carry them out. In this way, policies will be formulated with due consideration to their likely achievement.

Failure to consider appropriate planning and resourcing requirements when developing health and safety policies will lead to:

- Unrealistic planning, organization and staff deployment which places employees in compromising positions when implementing procedures.
- Inadequate resources being assigned to operations.
- Inappropriate use of insufficiently trained staff.
- Management system failures through inadequate control and lack of contingency planning.
- Little improvement and little development to procedures and practices.

## Health and safety policy links with other management systems

Health and safety policies and the procedures they underpin can form an integral part of policy making in other organizational areas. Quality management is one area where health and safety is frequently considered. This is particularly so where the organization holds a certificate of approval for a quality management system meeting system standards, for example BS EN ISO 9000: 1994 (BSI, 1994a; ISO, 1994a). Environmental management is a further organizational aspect where health and safety policies may be considered as part of the system procedures for addressing environmental safeguard of the workforce in conducting its business (BSI, 1994b; ISO, 1994b).

Any 'parent' management system will be structured to provide a framework for the management of the company and the satisfaction of its quality, safety, environmental and contractual obligations, while facilitating the optimum control of its business. Within a construction organization this system would be reflected in various levels of documentation: the system manual; sets of management procedures; and working instructions. The level of detail addressed increases from the manual to the procedures to the working instructions. These system elements would be translated into a project plan for health and safety for each specific construction project.

There are many similarities between the systems approach to effective health and safety management and quality management. While particular elements of any systems approach will vary according to specific system needs, the general concepts and principles of effective health and safety management and successful quality management are broadly the same.

There is wide acceptance within the construction industry that a well considered approach to quality management is a prerequisite to a successful organization. Those organizations which have recognized and responded dynamically to the need for better construction quality have undoubtedly benefitted in many direct and indirect ways. It is likely that those organizations who have extended their quality ethos to health and safety management will have experienced good standards of performance with fewer accidents. Often, this will have been facilitated by making the health and safety system an intrinsic part of their quality assurance (QA) procedures or total quality management (TQM) system. It is therefore appropriate that health and safety management policies be developed either alongside or as a part of quality policy making.

## Systematic approach

Effective policies and their implementation will be underpinned by the systematic identification, monitoring and control of health and safety risk. While the systematic approach is essentially governed by the developed procedures rather than the policy *per se*, policy making should be sufficiently explicit to lay the foundation for a formalized health and safety system to be established. Certainly, the extent to which health and safety consideration is reflected in business policy decision making is an important influence upon the system's ultimate effectiveness in practice.

**Review**
A real benefit to adopting a systems approach to health and safety management is that continuous improvement is inherently encouraged. As with any aspect of organization, learning from experience is essential to its overall health and vitality. Failure to capitalize on experience often leads to high-risk situations developing and the likelihood that no suitable strategy or procedures will be in place to expeditiously deal with them. Experience should be an influence on both the evolution of policies and the development of improved procedures. To achieve this, policies should make explicit the need to undertake a periodic review of health and safety performance against set plans and to audit the system used. A dynamic system will constantly measure, audit and review its organizational and project performance to improve its standards of practice.

**Statement of policy**
In many organizations, the health and safety policy will address a wide range of company interests and intentions. The aspects highlighted in a health and safety policy will depend upon the nature and business of the organization. Therefore, while meeting statutory obligations, policies can be as broad or as narrow as the organization considers appropriate.

As a health and safety policy is essentially the public face of an organization's recognition of and commitment to health and safety matters it should be declared in a written policy statement. This statement should be:

- unambiguous and presented in a clear format
- published with corporate identity
- amenable to public, inter-organizational and intra-organizational exposure.

In addition, the statement of policy may be:

- sufficiently flexible for incorporation in company publicity material such as its annual reports and marketing literature
- linked directly to other corporate aims, objectives and goals (it is not uncommon to refer to health and safety in conjunction with corporate statements on quality and environmental matters).

## The importance of the health and safety policy

The statement of health and safety policy should be formal in presentation, endorsed by the most senior executive of the company and be corporate in format. The statement should be a composite part of policy documentation and feature prominently in the company's health and safety manual and other associated manuals. The company's health and safety policies and the statement within which they are contained are absolutely essential to effective health and safety management since they establish the vision, culture and commitment throughout the entire organization and its personnel.

## Key points

Construction organizations which maintain a high standard of health and safety practices develop policies which:

- Clearly define the organization's corporate philosophy towards health and safety matters, in the context of its business activities.
- Place 'people' at the focus of health and safety policy development.
- Foster the development and maintenance of a supportive organizational culture where good health and safety practices become intrinsically the way of doing things.
- Generate commitment throughout the organization led by the company's executive management.
- Underpin effective health and safety planning and provide adequate resources to implement effective procedures.
- Consider links in health and safety policy with other aspects of organizational policy making, for example, total quality management (TQM) policies and procedures.
- Adopt a systematic approach to identifying, monitoring and controlling health and safety hazards and risk.
- Support a dynamic system of health and safety management where performance measurement, audit and review improves standards of practice and feeds back into evolving policies.
- Provide a clear and unambiguous statement of policy for use inside and outwith the organization.

## References

British Standards Institution (BSI) (1994a) *BS5750: Specification for Quality Systems*, BSI, London. (First published 1979, revised 1987, 1994 part of ISO 9000 series of standards.)

British Standards Institution (BSI) (1994b) *BS7750: Specification for Environmental Management Systems*, BSI, London.

Health and Safety Executive (HSE) (1993) *Successful Health and Safety Management*, HMSO, London. (First published 1991.)

International Standards Organisation (ISO) (1994a) *ISO 9000 Series: Specification for Quality Systems*, HMSO, London.

International Standards Organisation (ISO) (1994b) *ISO 14001: Specification for Environmental Management Systems*, HMSO, London.

# 15 Organization

## Introduction

Within the scope of responsibilities set by the CDM Regulations (HSE, 1994) the principal contractor is charged with delivering a safe working environment. Prerequisite to this is the establishment within the company's framework of the corporate organization and the project organization. These must effectively be brought together to ensure that the health and safety management subsystem is translated from policies at corporate level into procedures, instructions and working practices at project site level. Vital to success are the roles of the corporate health and safety manager/adviser, the health and safety supervisor based on site, first-line supervisors and all construction operatives. Organization for health and safety management focuses on delivering safe working practices and upon developing proactive leadership among health and safety managers and supervisors. This chapter presents an insight into the key elements of organization which supports these needs.

## The principal contractor's organization for health and safety management

The principal contractor's organization for health and safety management involves establishing the responsibilities and interpersonal relationships necessary to encourage a positive health and safety culture on site. The organization adopted for health and safety management has a bearing on the liaison between the participants and in particular upon the effectiveness of supervision of work gangs and the safe working procedures they put into practice (see Figure 15.1).

## Health and safety advice and supervision

For the management structure and the health and safety management subsystem described in Chapter 12 to be successful there must be effective

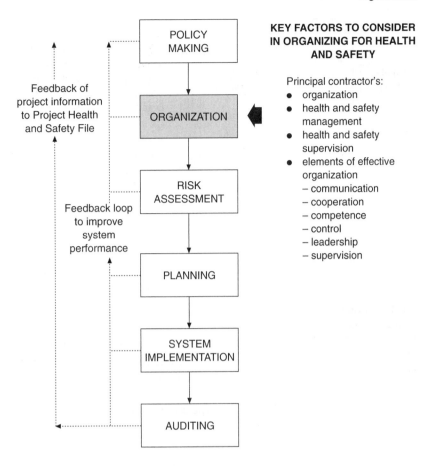

**Fig. 15.1** Key factors to consider in organizing for health and safety

implementation. A prerequisite to this is appropriate health and safety advice from the corporate organization to the project organization and diligent supervision on site. Only in this way will the health and safety policy of the organization be met and a safe working environment for project participants be delivered.

## The role of the corporate health and safety manager/adviser

The corporate health and safety manager/adviser has a multi-function role (see Figure 15.2) that encompasses tasks within the following dimensions of the organization.

1. The corporate organization.
2. The project site organization.
3. External to the organization.

Health and safety managers advise on and administer the following:

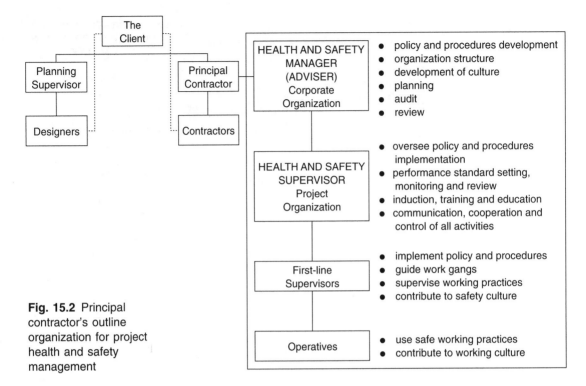

**Fig. 15.2** Principal contractor's outline organization for project health and safety management

## Within the corporate organization

- Health and safety policy development and implementation.
- Organizational structure for health and safety management.
- The development of the necessary supportive culture to ensure successful implementation of policies.
- Carrying out project risk assessment.
- Health and safety planning and its implementation.
- Health and safety induction and training for all employees.
- The development and implementation of monitoring mechanisms for performance on site.
- The implementing of efficient accident and incident investigation, reporting and review mechanisms.
- Review of safety performance and auditing of the management subsystems used for update and improvement.
- Advice on health and safety legislation relevant to the organization's business and its projects.

## Within the project organization

- The necessary support to site safety supervisors to ensure that all of the above aspects are effectively implemented on site.

- Effective liaison between the corporate parent management system and the health and safety management subsystem implemented on site.
- The maintenance audit and review of project health and safety performance including the performance of the site safety supervisor.
- Effective liaison between the various parties on site who have an input to health and safety management across the project.
- The adequate induction and training of all employees who are assigned to the project site.
- Appropriate cooperation with the client to ensure that all risk areas are identified and communicated to project participants.

### External to the organization

- Health and safety matters involving outside organizations including government departments, local authorities, environmental health officials, regulatory bodies, insurance companies, fire services, and the police.
- Liaison with the public in matters of public involvement, and safety on construction projects.
- Liaison with the media concerning matters of health and safety.

## The role of the project health and safety supervisor

The project health and safety supervisor's role is to ensure that all the aspects determined by the corporate health and safety manager/adviser are effectively implemented (see Figure 15.2). Health and safety supervision has two important aspects:

1. Safe practice.
2. Leadership.

### Safe practice

A fundamental task of the health and safety supervisor is to ensure the use of safe working practices. These must be in accordance with safety objectives and to meet set performance standards. This involves the supervisor in providing the necessary instruction, direction, guidance and help to first-line supervisors and operatives that makes tasks safe to perform and where a sensible approach is taken. To achieve this it is essential that operatives are aware of the potential risks when undertaking any task and consciously think about and apply safe practice to the task. The key elements of developing safe practice are effective:

1. Communication.
2. Cooperation.
3. Competence.
4. Control.

The effective management of these elements is at the heart of successful supervision. They are discussed subsequently.

### Leadership

Proactive leadership is absolutely essential to effective health and safety management. In order to encourage first-line managers and operatives to work together towards creating a safe working environment there must be effective leadership.

The site safety supervisor must set and be seen to set an exemplary example as a model to the workforce. An aspect of particular importance is the need to make employees aware of the risks and then to manage those risks in the context of their own work practices and the practices of their colleagues. The ultimate aim is to elicit the support and participation of the total workforce and all persons operating on site. This can only be achieved through effective supervision.

## Key elements of effective organization

There are four key elements to effective organization for health and safety management.

### Communication and information

Efficient and effective communication is an essential element in any organization and when implementing any management concept. Communication involves three primary routes of information flow:

1. Into the organization.
2. Within the organization.
3. From the organization.

Communication into the organization is concerned with an influx of information on health and safety matters which informs development of policies, standards and managerial practices. Such information includes updated legislation, technical advances and variations in health and safety management approach.

Communication within the organization is crucial if the workforce is to understand and implement the company's health and safety policy successfully. Communication systems are important to such success and focus on: developing greater awareness for policy; underpinning policy by beliefs which drive support; demonstrating commitment; sharing plans, procedures and expected practices; and seeking improvements in performance. Organizational systems to support effective communication involve both formal and informal mechanisms and flow up, down and across the structure of both the corporate and project organizations. In all its forms, internal communication serves to share the organization's commitment toward monitoring a strong positive health and safety culture. First-line managers play an important role here as they are the intermediary between the safety supervisor and the general workforce.

Communication flow from the organization is a less obvious but important aspect. Many health and safety reporting practices demand communication to and liaison with external bodies. A prominent example is the Health and Safety Executive. A further example might be rapid communication mechanisms to handle emergency situations on site. This may involve liaison with local emergency services.

### Cooperation between all participants

The emphasis given to participation in health and safety management by the CDM Regulations (HSE, 1994) makes cooperation between everyone involved with a construction project absolutely essential. Cooperation between the various contractual parties and between employees within the parties is influential to the successful safe outcome of the project.

Within dynamic principal constructing organizations employees at all levels will have the opportunity to be involved with developing management procedures and working instructions, determining performance standards and reviewing performance.

Involvement is the key attribute to encouraging ownership of health and safety policies, sharing objectives and enacting safe working practices. Ways in which cooperation can be encouraged include the development of health and safety committees, suggestion schemes, safe practice awards and brainstorming health and safety circles where collective discussion informs practice, review and future policy development.

### Competence of managers and employees

A prerequisite to meeting the requirements of the CDM Regulations is to ensure the competence of all employees involved with the project. Managers and first-line supervisors must be aware of current health and safety legislation while operatives must be able to implement safe working practices, not only for their own safety but the safety of others. Key to this are the effective management of recruitment, instruction and training.

It is important that the health and safety supervisor ensures that:

- Recruitment effectively identifies staff and operatives with the necessary levels of competence, and awareness of and commitment to health and safety practices.
- Training needs are identified and followed up with a structured programme of education and training to maintain current competences.
- Induction and instruction to health and safety practices are effectively administered to reinforce competent practice as soon as an employee starts on site.
- Competence gaps are identified and responded to as and when they occur, for example as a result of staff absence or transfer.

Effective training is vital to maintaining a competent workforce. Training will assist employees to develop the awareness, skills, attitudes and knowledge to make them competent in utilizing safe working practices. This can include on-the-job instruction, off-the-job education and training, mentoring and guided application.

In general, there are three main types of training needs:

1. Organizational.
2. Job related.
3. Individual.

Organizational needs centre on awareness and education concerning health and safety policies, organizational structure, roles and responsibilities, gaining support, communication, and implementation of the health and safety management procedures. Job related needs focus on management aspects and workforce aspects and include leadership, communication, management practice, risk assessment and mitigation measures and management review. Individual needs are specific to the individual and vary according to the situation and circumstances of the person. Needs might include induction for new employees or re-training in new procedures.

### Control of site activities

Developing and maintaining effective control is again a prerequisite to all organizations and core to the effective management of many organizational aspects including health and safety. Health and safety managers, first-line supervisors and general site management staff are responsible for controlling all health and safety matters on site. The three key elements of effective control are:

1. Planning.
2. Monitoring.
3. Review.

In association with corporate health and safety management a number of functions must be implemented by the project health and safety supervisor to ensure effective control:

- The development of plans which reflect the organization's policy and objectives towards health and safety management.
- The proactive and vigilant monitoring of site activities to ensure that plans are effectively implemented.
- The review of health and safety performance in relation to implemented plans, highlighting difficulties and the effects of actions taken in mitigation.

It is fundamental that first-line work supervisors, responsible for health and safety performance, clearly appreciate just what they are responsible for and for what they are accountable. Performance standards are the bridge between performance expectation and responsibility achievement. Performance standards will encompass who is responsible for each health and safety activity, the tasks that must be undertaken, when tasks must be undertaken, and the expected outputs of the activity.

Linking performance to responsibility is a function of the supporting mechanisms to practising managers developed within an effective health and safety management subsystem. There should be clear reference to health and safety responsibility within job descriptions, formal performance review and appraisal, and unequivocal measures of management action when health and safety difficulties occur.


## Effective supervision

The key elements of effective organization for health and safety management on site are brought together by good supervision. Supervision actively gels the positive attributes of communication, cooperation, competence and control ensuring that each makes a solid contribution to the subsystem's holistic implementation. Effective supervision can encourage greater awareness, pursuit of common policies, objectives and goals, open communication and understanding, better leadership and greater support and participation in health and safety management.

The effectiveness of supervision is demonstrated where it is proactive rather than reactive. For this reason the organization of the supervisory role must focus upon active health and safety controls. This requires that not only are management procedures and working instructions employed rigorously but moreover that first-line supervisors think intuitively about their implementation.

Leading by example is paramount in delivering effective supervision and developing the holistic health and safety culture that the project organization requires. Communicating the right signals to employees is highly significant. Managers and supervisors should be cognizant that they have a vital role to play in supporting positive health and safety behaviour and practices both throughout the organization and on site.

## Key points

In this chapter it has been identified that:

- The principal contractor's organization for health and safety management involves establishing within the management subsystem the responsibilities and interpersonal relationships necessary to encourage a positive health and safety culture.
- The health and safety manager/adviser, a position within the corporate organization, has a multi-function role that encompasses the corporate organization, the project site organization and the external dimension.
- The health and safety supervisor, a position within the project organization, also has a multi-function role that encompasses the implementation of safe working practices and leadership of site employees.
- Key elements to effective organization for health and safety management are communication and information, cooperation between all participants, competence of managers and employees, and control of site activities.
- Effective organization for health and safety centres on effective first-line supervision and this is influenced to a great extent by good proactive and visionary leadership.
- Effective supervision centres on the provision to all employees of instruction, guidance, mentoring, training and above all support for and mutual reinforcement of safe working practices.

## Reference

Health and Safety Executive (HSE) (1994) *The Construction (Design and Management) Regulations 1994*, HMSO, London.

# 16 Risk assessment

## Introduction

This chapter presents an overview of the process of risk assessment. The CDM Regulations (HSE, 1994) emphasize risk assessment during the design phase and place clear responsibilities upon the designer and the principal contractor. Risks identified during the conceptual development of the project will allow the designer to consider their likely effects and take mitigating steps. This is undertaken through detailed health and safety risk assessment during design review. This information forms an important part of the pre-tender health and safety plan. Hazards and risks identified in the designer's risk assessment will be followed up by the principal contractor when compiling the construction health and safety plan, a further requirement of the CDM Regulations.

## Risk assessment technique

Risk assessment is a well-established and recognized analytical technique used widely across many different fields of business, commerce and industry. In the context of health and safety within construction its application focuses upon the identification, assessment and control of risk to minimize those hazards which can arise in the course of undertaking a project. The key factors for the principal contractor to consider are shown in Figure 16.1.

The CDM Regulations emphasize risk assessment within the duties and responsibilities of both the designer and principal contractor. For example, a designer is required to avoid foreseeable risks to health and safety of any person carrying out construction work, to combat risks at source, and to give priority to measures which protect the whole workforce.

Risk assessment for construction health and safety involves three key activities:

1. Hazard identification.
2. Evaluation of risk.
3. Prevention and protection measures.

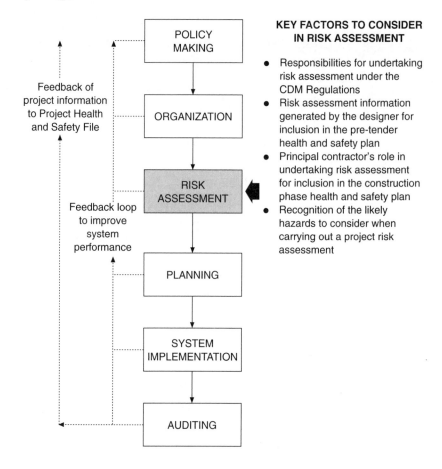

**KEY FACTORS TO CONSIDER IN RISK ASSESSMENT**

- Responsibilities for undertaking risk assessment under the CDM Regulations
- Risk assessment information generated by the designer for inclusion in the pre-tender health and safety plan
- Principal contractor's role in undertaking risk assessment for inclusion in the construction phase health and safety plan
- Recognition of the likely hazards to consider when carrying out a project risk assessment

**Fig. 16.1** Key factors to consider in risk assessment

### 1. Hazard identification

A hazard is something which presents a potential to cause harm. This could be through the occurrence of an accident or exposure to a dangerous situation, material or substance. Within the construction industry, ever-present hazards leading to fatal and serious injury commonly involve working at heights, use of ladders and scaffolds, collapse of temporary structures, use of vehicles, mechanical plant and equipment, and exposure to harmful substances. Further details are given in Chapter 2. Hazard identification involves the systematic recognition of any aspects of a project which have a potential to be a danger to those persons working on or being around that project.

### 2. Evaluation of risk

Risk is 'likelihood that a specified undesired event will occur due to the realization of a hazard' (Croner, 1994). Once a hazard has been identified the degree of risk must be determined. Two factors are influential to this determination:

1. The *severity of harm* – the level of harm that a circumstance would create and

2. The *likelihood of occurrence* – the frequency of a hazardous circumstance.

The evaluation of risk can be made with the aid of a simple calculation:

severity of harm $\times$ likelihood of occurrence = degree of risk

A risk assessment gives the statistical probability of a hazardous event occurring. The outcome is based on a body of information, both qualitative and quantitative, obtained from factual experience to develop a numerical figure which represents the degree of risk. The following example illustrates this.

Suppose a situation indicates:

the *severity of harm* is 3

and

a *likelihood of occurrence* is 4

where:

the severity of harm is an assigned value on a six-point scale from minor injury (1) to death (6), based on factual information (see Table 16.1)

and

the likelihood of occurrence is an assigned value on a six-point scale from remote (1) to highly probable (6), based on factual information (see Table 16.2)

The degree of risk is:

severity of harm $\times$ likelihood of occurrence

therefore

$3 \times 4 = 12$ (degree of risk)

**Table 16.1** Evaluation criteria for severity of harm

| *Project A1* | *Evaluation criteria for hazard severity* |
| --- | --- |
| *Assigned value* | *Description* |
| 1 | Minor injury – no first aid attention |
| 2 | Illness – chronic injury |
| 3 | Accident – needing first aid attention |
| 4 | Reportable injury – under RIDDOR* |
| 5 | Major injury – under RIDDOR* |
| 6 | Death |

*RIDDOR: Reporting of Injuries, Diseases and Dangerous Occurrences Regulations 1995

**Table 16.2** Evaluation criteria for likelihood of occurrence

| Project A1 | Evaluation criteria for likelihood of occurrence |
|---|---|
| Assigned value | Description |
| 1 | Remote – almost certain not to occur |
| 2 | Unlikely – occurrence in exceptional circumstances |
| 3 | Possible – certain circumstances would influence occurrence |
| 4 | Likely – could ordinarily occur |
| 5 | Probably – high chance of occurrence |
| 6 | Highly probable – 100% chance of occurrence |

The degree of risk (12 in the above example) is a numerical value which is a proportion of the possible maximum degree of risk. The maximum risk is 36 (the severity of harm on the six-point scale, multiplied by the likelihood of occurrence on the six-point scale, see Tables 16.1 and 16.2). The value (12) can, perhaps, be more meaningfully expressed as a percentage of the maximum risk (36), i.e. 33 per cent. This is the percentage chance of the hazardous event occurring.

In risk assessment, rather than expressing the degree of risk in percentage terms, a *priority rating* is given, for example:

- Low Priority (L)
- Medium Priority (M)
- High Priority (H)

where, in the example, valued criteria give the following priority bands:

- (L) = 3–9%
- (M) = 10–44%
- (H) = 45–100%

However, bands on the points scale can vary according to the value assigned to the criteria. Therefore, in the example, the risk of 33 per cent would lie in the medium priority (M) band.

An alternative approach is to grade risk on a scale of 1 to 10, where 1 = 1% and 10 = 100%. A table can be developed to determine the degree of risk where the severity of harm and likelihood of occurrence have been assigned priority ratings: low, medium and high (see Table 16.3).

The evaluation of risk enables the value to be calculated for any hazard identified. The greater the value, the higher the priority and therefore the more thought that should be given to, and effort that should be placed upon, avoiding or managing the risk.

**Table 16.3** Determination of priority rating for risk

| Project A1 | Priority rating of risk | | |
| --- | --- | --- | --- |
| | Likelihood of occurrence | | |
| Severity of harm | High/low 10% | High/medium 50% | High/high 100% |
| | Medium/low 5% | Medium/medium 25% | Medium/high 50% |
| | Low/low 1% | Low/medium 5% | Low/high 10% |

### 3. Prevention and protection measures

When an evaluation of risk has been considered, the principles of prevention and protection should be applied. . . . The principles, in summary are to: a) avoid risk; b) combat risk at source; c) control risk.

(Croner, 1994)

For example, a requirement of the CDM Regulations is for the designer to identify risks associated with the construction work and to redesign to avoid those risks occurring.

Where avoiding risk proves impossible, the designer must make provision to combat the risk and ensure that appropriate information is passed to the planning supervisor and thereby to the principal contractor. This will allow safe systems of work to be developed which will control the level of risk from those hazards identified.

## The designer's role

Within construction, the design process progresses through three main development stages, these are:

1. Project feasibility and outline concept.
2. Scheme design and layout.
3. Detail design and specification.

The most effective time to consider project health and safety and eliminate potential hazards is early in the design process, during project feasibility and outline conceptual development. As any construction project progresses the opportunity to design-out the hazards will diminish. Risks identified during conceptual development will allow the designer time and space within the design development schedule to determine appropriate mitigation measures.

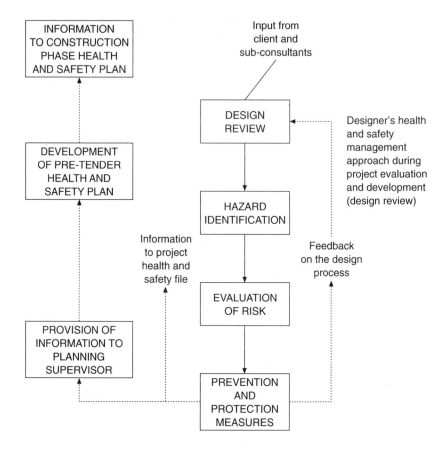

**Fig. 16.2** Designer's contribution to health and safety management during project evaluation and development (design review)

Risks identified during the detail design stage will likely have no such latitude. The best that might be achieved in the later stages of the design process is that potential control measures can be suggested.

It is imperative that the key technique of risk assessment is utilized from the earliest stage of the design process. As the design evolves throughout the various stages of the design process so risk assessment is an evolving process. As a project's design develops it will be reviewed periodically to ensure that it meets the client's brief and the designer's intentions. Design review provides an excellent opportunity to evaluate the project's health and safety dimension. It forms the basis of the designer's contribution to project health and safety management.

**Design review** Figure 16.2 illustrates the fundamental aspects of the designer's role in hazard identification and risk assessment. The approach is based upon developing the project design within a management information loop. The system commences with the receipt of information necessary to developing the design from the client and any contributing sub-consultant. It concludes in the information necessary to take forward the pre-tender health and safety plan to tendering and the formulation of the construction phase health and safety plan. Information can also be directed to the project health and safety file.

Health and safety design review involves three key procedures, commensurate with the three key activities involved in the risk assessment process:

1. *The identification of hazards* – the designer should identify each hazard within the design. Each should be given a reference number and be clearly described.
2. *The evaluation of risk* – the designer should carry out an assessment of the potential risks associated with each hazard identified (this is achieved using the risk assessment strategy and procedures described previously).
3. *The consideration of prevention and protection measures* – the designer should, where possible, take action immediately in response to the identified hazard. This might be achieved through a redesign of the hazardous element. However, some hazards may not be dealt with at the time of identification and the designer will need to revisit the issues as the design evolves and more information becomes available. Certain hazards may be recognized and recorded but left unresolved for the contractor to address. This would be the case where, for example, the hazard was posed not by the design element but through the methods to be used to carry out the works.

**Hazard identification and risk assessment records**

Systematic gathering and recording of information concerning project risk is essential to the designer's health and safety design review. The documentation necessary to establish a systematic approach to health and safety design review and to meet the requirements of the CDM Regulations are as follows:

- the designer's hazard identification record
- the designer's risk assessment record.

### The designer's hazard identification record

Figure 16.3 illustrates a pro forma document which can be used to record information for the designer's hazard identification. The information recorded should include the following:

- *Project; Designer; Stage of Work; Date* – This information identifies the record, relating it to the particular construction project. In addition, the date and stage of work are recorded, thereby relating the record to changes that take place as the project progresses.
- *Reference number* – As each hazard is identified by the designer when formulating the design, it should be given a reference number. This is to ensure that each hazard can be traced throughout the design process as the various design reviews are carried out.

| DESIGNER'S HAZARD IDENTIFICATION | | | |
|---|---|---|---|
| PROJECT: | | | |
| DESIGNER: | | | |
| STAGE OF WORK: | | DATE: | |
| | | | |
| REF NUMBER | DESCRIPTION OF HAZARD | ACTION REQUIRED | PERSONS TO INFORM |
| | | | |

NOTES:
For example – drawing reference
                    – reference to specifications

PERSONS TO INFORM – LEGEND:
PS – Planning Supervisor
SE – Structural Engineer
ME – Mechanical & Electrical Eng
QS – Quantity Surveyor
LA – Landscape Architect
PC – Principal Contractor
R – Review later

**Fig. 16.3** Designer's hazard identification record: suggested pro forma

- *Description of hazard* – Each hazard identified should be briefly described.
- *Action required* – Details of the action taken at the design review to eliminate or reduce the hazard should be described. If the hazard is to be the subject of a later review or the matter to be passed to the principal contractor for attention, then this should be noted.
- *Persons to inform* – Information on hazards identified will invariably need to be passed to other participants to the project, for example, the planning supervisor, sub-consultants and the principal contractor. Any persons to inform should be recorded in this section. To aid this a legend specifying the initials of the various parties can be given at the foot of the pro forma.
- *Notes* – Any other information, for example references to drawings and specifications, should be described in the notes section.

### The designer's risk assessment record

Figure 16.4 presents a pro forma document which can be applied to recording information for the designer's risk assessment record. This form records information which allows judgements to be made about the degree of risk and what actions may be taken to reduce the risk. The information recorded should include the following:

- *Project; Designer; Date; Sheet number* – As with the project identification record, the risk assessment record should give basic information relating the record to the particular project, designer and date of compilation.
- *Reference number* – This is the reference number which was allocated to the hazard as recorded in the hazard identification record.
- *Element of work* – The element of the work, or activity, should be briefly described, for example, excavation for basement of main office building.
- *Potential hazard/risk* – Aspects of the design posing the potential risk should be described, for example, working in close proximity to electricity cables.
- *Persons at risk* – As different controls may be required for particular persons, these persons should be listed on the record, for example, the public, site operatives and visitors to site.
- *Risk rating* – An estimate of the level of the risk should be given. The method of determining the degree of risk – the severity of harm and the likelihood of occurrence – is described on pp. 156–9.
- *Action at design stage* – Actions taken to eliminate or resolve the degree of risk by altering the design or recommendations to the principal contractor for inclusion in the contractor's method statement should be described.
- *Action taken* – Where any actions are taken the persons responsible for initiating the actions should be noted and the date when the actions are taken noted.
- *Risk control possibilities* – The designer may identify risks which require a particular management course and for which specific action must be taken by others – for example a particular construction method and sequence.

| DESIGNER'S RISK ASSESSMENT | | | | | | | | | | |
|---|---|---|---|---|---|---|---|---|---|---|
| DESIGNER: | | | PROJECT: | | | | DATE: | | SHEET OF | |
| REF NUMBER | ELEMENT OF WORK | POTENTIAL HAZARD/RISK | PERSONS AT RISK | RISK RATING | | | ACTION AT DESIGN STAGE | ACTION | | RISK CONTROL POSSIBILITIES |
| | | | | L | S | R | | BY | WHEN | |
| | | | | | | | | | | |
| | | | | | | | | | | |
| | | | | | | | | | | |

RISK RATINGS:        L – Likelihood (medium/high);        S – Severity (low/medium/high);        R – Risk (severity × likelihood)

**Fig. 16.4** Designer's risk assessment record: suggested pro forma

Information should be given in this column for the attention of the principal contractor.

**The provision of information**

The designer is duty bound by the CDM Regulations to avoid foreseeable risks, combat risk at source, protect the entire workforce and communicate appropriately on any risk to the project as a result of the design solution.

It is essential that the designer identifies all the potential hazards within the design, together with those that may occur during construction as a result of the design. Hazards must not only be recognized but highlighted to other contractual parties and be priced for by tendering prospective principal contractors, subcontractors and suppliers.

The information collected by the designer during design review, presented in the hazard identification record and risk assessment record (see Figures 16.3 and 16.4), will be passed to the planning supervisor. It will be included in the pre-tender health and safety plan and ultimately form part of the project health and safety file. The timely and effective provision of information throughout the design process is essential in providing accurate information for the pre-tender health and safety plan. Good information is vital because the information will be used in so many subsequent health and safety management stages – the construction health and safety plan, health and safety management on site and the generation of the health and safety file.

In addition, closing the system loop is imperative to the designer because much can be gleaned about the design process within the project situation for reference on other projects.

## The principal contractor's role

In meeting the requirements of the CDM Regulations, every contractor to the project must prepare a risk assessment. This will address all potential hazards on the project and identify the population at risk, such as employees, visitors to site, and the public. The principal contractor must prepare a risk assessment in which the risk assessments of any other participating contractor must be reflected. The risk assessment will also follow up any hazards identified in the designer's risk assessment which were highlighted for the principal contractor's attention.

Risk assessments prepared by principal contractors and contractors often comprise two parts. The first part reflects the hazards and risks associated with the normal activities that they will undertake in many construction projects and is reflected in a generic section to the risk assessment. The second part will be tailored to accommodate hazards and risks specific to the project.

Having identified potential hazards which are generic to normal activities or specific to the project, the population potentially at risk must be identified, the degree of risk assessed and methods of mitigation considered. Each hazard must be considered in turn and appropriate documentation compiled for inclusion in the construction health and safety plan. A suggested pro forma record is shown in Figure 16.5. Information from the principal contractor's risk assessment is used to develop specific safe working procedures for implementation on site and these augment the generic safe working procedures contained in company manuals of procedures and files of working instructions.

**Project hazards: activities with a risk to health and safety**

The following lists suggest potential hazards which might be given consideration as part of a principal contractor's hazard identification and risk assessment. Aspects reflect hazardous situations relating to:

- the existing environment
- use of equipment.
- working with materials and substances
- implementation of procedures
- site wide influences
- particular processes involved
- works governed by associated Regulations

**Working at heights**

- openings in floors
- slippery surfaces
- gaps on working platforms
- stairwells and lifts
- overhead power lines
- elevated work platforms
- safety harnesses and lifelines
- guardrails and toe boards
- scaffolds
- ladders and hoists
- safety below working at heights
- safety screens

**Use of electricity**

- faulty insulation
- presence of water
- humidity

## Principal Contractor's Risk Assessment

| Project: | | | | | Document Ref No: | |
| Contractor: | | | | | Specialist Discipline: | |
| Assessor: | | | Signed: | | Date: | |
| Activity/Element | Potential Hazards | Population at Risk | Risk Rating | | Priority | Control Measures Specified |
| | | | L | S | R | | |

Sources of information:

Legend:
L – Likelihood
S – Severity
R – Risk (Severity × Likelihood)

**Fig. 16.5** Principal contractor's risk assessment: suggested pro forma

167

- conductive clothing
- power cable management
- underground cables
- power cables at heights
- circuit breakers
- transformers

### Fire

- flamable dust and vapours
- spillage
- hot-work activities, e.g. welding
- combustible materials
- rubbish removal

### Plant and equipment

- appropriate tool choice
- fault inspectors
- site movement
- safe speeds
- alert to use and movement
- supervision

### Hazardous materials

- chemicals
- petrol
- paints
- adhesives
- storage
- warning signs

### Excavations

- trench collapse
- falling objects and materials
- nearby objects and obstructions
- safe digging practices
- excavated spoils near trenches
- services
- slides and cave-in
- inspections

- earthwork retainment
- warning signs

## Confined spaces

- gas and vapours
- liquids
- shafts and sewers
- services

## Manual handling

- safe lifting practice
- aids to move loads
- team lifting of heavy objects

## Personal protective equipment (PPE)

- helmets
- high visibility clothing
- boots
- goggles
- gloves
- harnesses
- ear defenders
- face masks

## Site housekeeping

- workplace tidiness
- safe accesses and workplaces
- tool storage
- rubbish removal

## Natural and personal elements

- wind
- sun
- rain
- cold and heat extremes
- fluid intake
- physical distress
- sensible work limits

Site aspects and public safety

- signs
- barriers
- walking areas
- falling objects
- traffic management
- parking
- access/egress
- unauthorized access

## Key points

This chapter has identified that:

- The CDM Regulations (HSE, 1994) emphasize risk assessment within the responsibilities of the designer, principal contractor and contractors.
- Risk assessment involves three key activities: (i) hazard identification; (ii) evaluation of risk; and (iii) determination of prevention and protection measures.
- The designer has a crucial role to play in hazard identification and risk assessment during design review where redesign may help to alleviate the risks at source.
- Design review information will be passed to the planning supervisor for inclusion in the pre-tender health and safety plan. Some aspects will be highlighted for the attention of the tendering principal contractor.
- Risk assessment will be undertaken by the principal contractor and all other contractors to the project. This will form part of the tender submission in the form of the construction health and safety plan.

## References

Croner (1994) *Croner's Management of Construction Safety*, Croner Publications Ltd., Kingston upon Thames, Surrey.

Health and Safety Executive (HSE) (1994) *The Construction (Design and Management) Regulations 1994*, HMSO, London.

# 17 Planning

## Introduction

This chapter focuses on planning for health and safety. The CDM Regulations (HSE, 1994a) necessitate a two-part approach to planning with the objective of delivering a health and safety plan for the project. The first part is the responsibility of the planning supervisor, working with the designer, who must produce a pre-tender health and safety plan. The second part is the responsibility of the appointed principal contractor, who builds upon the pre-tender plan and considers the construction phase health and safety planning issues. The health and safety plan is, therefore, a mechanism for bringing the project participants together to improve communication and teamwork on matters of project safety, health and welfare. The focus of the plan is to identify the potential hazards to health and dangers to safety through each stage of the construction process and assess their degree of risk. This consideration forms the basis for developing the safe working practices which are essential to the on-site construction works.

## Planning for health and safety

The CDM Regulations are directly concerned with planning for the encouragement of teamwork in and coordination of the management of project health and safety. The Regulations recognize the complexities characteristic to the construction processes. In addition, they acknowledge the traditional separation of the participants that come together to form the project team, together with the challenges and problems that this can bring.

The essential aim of the CDM Regulations is therefore, to ensure that health and safety is consciously considered and planned for by all project participants. Furthermore, to ensure that health and safety becomes intrinsic to the management of the construction project from the outset.

## The health and safety plan

The purpose of the health and safety plan is to document all information relevant to project health and safety and ensure that this information is available to those associated with the project who need it. The health and safety plan is dynamic and develops with the project. It is developed in two parts. The first part is associated with the design and planning of the works prior to tendering or contractor selection. The second part is associated with construction works on site. The two parts are therefore:

1. Development of a pre-tender health and safety plan by the client organization.
2. Development of a construction phase health and safety plan by the principal contractor.

These two planning documents, which roll on as an active health and safety plan throughout the total construction process, provide information for a third and equally important piece of project documentation – the health and safety

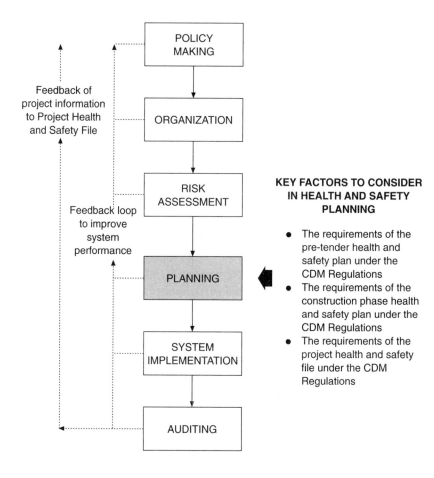

**KEY FACTORS TO CONSIDER IN HEALTH AND SAFETY PLANNING**

- The requirements of the pre-tender health and safety plan under the CDM Regulations
- The requirements of the construction phase health and safety plan under the CDM Regulations
- The requirements of the project health and safety file under the CDM Regulations

**Fig. 17.1** Key factors to consider in health and safety planning

file. The health and safety file is a project record of health and safety information for the client or end user. The file is retained by the client at the end of the project.

It is this fundamental requirement for systematic planning which forms the basis for a systems management approach within which risk assessment is the central theme. See Figure 17.1.

While the direction of this book is towards the role and responsibilities of the principal contractor and therefore the focus of planning is the construction phase health and safety plan, it is also essential to appreciate the thrust of the pre-tender health and safety plan. Indeed, the information gathered for the pre-tender plan is invaluable to the principal contractor when developing the construction phase plan. This chapter looks at both pre-tender and construction phase health and safety planning.

## The pre-tender health and safety plan

The pre-tender health and safety plan is the responsibility of the client and is produced by the planning supervisor. The pre-tender plan documents those health and safety issues that need to be considered when planning the site works by potential principal contractors when they are selected or tender for the contract.

The pre-tender health and safety plan should include:

- A general description of the project and works.
- The timescales for completion.
- Details of known health and safety risks to operatives.
- Information required by contractors for them to demonstrate that they are competent and have adequate resources.
- The information needed by the principal contractor to develop the construction phase health and safety plan.
- Information required by contractor on compliance with welfare provisions.

## The construction health and safety plan

For planning the construction phase the principal contractor develops the health and safety plan with consideration to those issues which impinge upon the health and safety management of the project.

The construction health and safety plan should include:

- The arrangements for ensuring health and safety of all persons who may be affected by the project.

- The arrangements for the management of health and safety of the project and monitoring of compliance with health and safety law.
- Information concerning welfare arrangements for the project.

The information that has been generated throughout both the pre-tender and construction phase health and safety planning is recorded in the project health and safety file.

## The health and safety file

> This is a record of information for the client/end user, which tells those who might be responsible for the structure in future of the risks that have to be managed during maintenance, repair or renovation.
>
> (HSE, 1994b)

The planning supervisor must ensure that the file is prepared as the project progresses and is handed to the client when the project is complete. The client must ensure that the file is available to persons who work on any future design, construction, maintenance and repair, and demolition of the structure.

The main parties to the project employed by the client, namely the design consultant and principal contractor, each play an essential part in developing health and safety management systems and working procedures (Figure 17.2). These identify, assess and control risk both within and across their professional boundaries. In addition, consistent with good planning for systems management, there are information feedback loops both within the span of control of the parties' activities and across the construction processes. In the context of the CDM Regulations this ensures a full contribution by each party to health and safety planning, the development of implementation systems and project review. Good planning will be evidenced through the delivery of the health and safety file.

## Developing the pre-tender health and safety plan

The CDM Regulations propound that the management of health and safety for any project commences with planning, design and specification. Clients and designers can, therefore, make very considerable contributions in identifying hazards and assessing and managing risks by formulating a clear and comprehensive health and safety plan early in the project development sequence. This was outlined in Chapter 16.

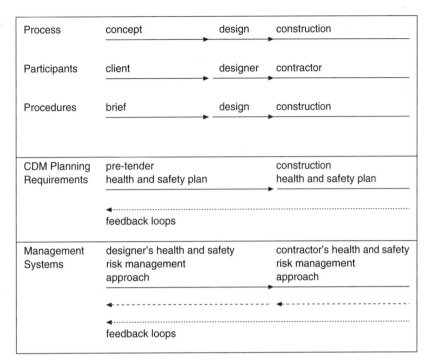

**Fig. 17.2** Management systems to accommodate project health and safety planning

The pre-tender health and safety plan is a vital key element in the identification and assessment of risk. The information gathered is essential to project evaluation and development. Furthermore, the information is prerequisite to developing the construction health and safety plan which lays the foundation for a safe working environment on site.

The pre-tender health and safety plan should contain information such that prospective principal contractors tendering for the project can plan and cost for safety measures in their price. Where a contract is awarded or negotiated rather than bid for, the same information should still be provided for this first stage of health and safety planning.

**Contents**  The information required for developing a pre-tender health and safety plan will be gathered within three broad aspects of the project; these are:

- the existing environment
- the design
- the site.

Information will be input by the client, designer and planning supervisor. The information is project specific and therefore will vary from one project to the next. Notwithstanding, information can be gathered under standard headings to simplify matters.

*Managing Construction for Health and Safety* (HSE, 1995a), the HSE approved code of practice, suggests nine headings under which information for the pre-tender health and safety plan might be gathered:

1. nature of the project
2. existing environment
3. existing drawings
4. design
5. materials
6. site elements
7. overlap with the client's activities
8. site rules
9. continuing liaison.

### Nature of the project

This section will contain such information as the name and address of the client, the location of the project site, details of the construction, and time-frame for the construction phase.

### Existing environment

This section will describe the environment on and around the project site. This is important to the consideration of site access and egress and to activities taking place at the interface between the site and its environs, for example materials delivery, or the movement of heavy mechanical plant. Details of existing service utilities, for example, water, electricity and telephone, together with possible concealed services such as gas pipes or water mains, should be included as these may represent a potential safety risk. Existing buildings and structures should also be included in the plan as these could pose a possible hazard.

### Existing drawings

Existing drawings may be available from landowners, building owners or occupiers, utility companies, or local authorities. These may record site layouts, structural details of existing or neighbouring properties and location of existing services. Any such drawings and specifications should be included in the plan. If the project involves work to recently completed existing build-ings or structures, there will be available health and safety files from which useful information might be drawn.

### Design

Findings of the designer's risk assessment should be included in the plan. Any hazardous works should be identified and described and requests for method statements highlighted for the attention of the principal contractor.

## Materials

Details of any potentially hazardous materials specified in the design and for which specific precautions will be required should be given. This relates to particular materials noted as hazardous by the designer, for example special paints or adhesives. Materials which are generally hazardous, for example cement or plaster, should be understood implicitly as such by any competent construction operative.

## Site elements

This section will include information which is important to any persons present on the site, for example, temporary accommodation, delivery, unloading and storage, pedestrian and vehicular traffic routes, services and amenities, and site boundaries.

## Overlap with the client's activities

Where the project involves work to existing premises there may be restrictions to working practices. For example, working hours may be limited, noise and pollution controls may be required beyond those usually encountered, or access may be limited to specific areas or times. Any such details should be included in the plan.

## Site rules

The client, designer or planning supervisor may impose rules to be observed on site, for example vehicular speed restrictions, or a no-smoking policy. These should be clearly stated in the plan.

## Continuing liaison

The plan should contain details of key contacts for the client, designer and planning supervisor. Information will include the full name, address and telephone number of each contact. In addition, the protocol for requesting particular information, for example instructions, should be described.

**Inputs to development**     ## The client

The client has an essential contribution to make to the pre-tender health and safety plan. This takes the form of providing information which is key to the project's development. Such information might include:

- location and description of the works
- time-frame for project development and construction
- existing documentation and drawings of the site
- details of site topography and conditions
- conditions of existing buildings and structures
- existence and location of temporary services

- arrangements for site access and traffic management
- need for security arrangements
- facilities for storage and protection
- potential restrictions to work routines and general operations
- findings of risk assessments.

These aspects represent some of the information that may be provided by the client and the consultants that they employ. Some information may be available from existing sources, while some might have to be acquired through undertaking specific surveys. The salient point is that all the information that is provided by the client is project site specific. Details relating to the construction form is incorporated in the information provided by the designer.

### The designer

The designer will be able to provide information to the planning supervisor considering both the site and the project. Such information may include the following:

- details of existing buildings internal layouts
- results from soil investigation and analysis
- phasing and sequencing of the works
- hazards presented by the proposed design.

## Developing the construction phase health and safety plan

Following the appointment of the principal contractor, the principal contractor assumes responsibility for the development of the construction phase health and safety plan. On some projects there will be project development and design works still being undertaken although the principal contractor has been appointed. In this situation, these works remain the responsibility of the planning supervisor until they are at the appropriate stage to be passed to the principal contractor for inclusion in the construction health and safety plan.

The information included in the pre-tender health and safety plan forms the basis of development for the construction health and safety plan. As stated previously, the client has a responsibility to ensure that the health and safety plan is prepared subject to any outstanding aspects under the control of the planning supervisor, before construction commences on site.

In compiling the construction health and safety plan, the principal contractor will provide information to enable:

- development of a framework for managing health and safety of all those involved in the construction stage of the project
- the contributions of other organizations involved in the construction phase to be included
- the organization of and action needed to investigate hazards, including informing all personnel who might be placed at risk
- good communication by the contractual parties
- the development of method statements, design details and specifications for all project contractors (subcontractors)
- the security of the site to ensure that only authorized personnel gain access.

**Contents**    The content and detail of the plan will depend upon the nature and characteristics of the project, the degree of risk associated with the site and the works and the amount of health and safety information encompassed within other project documentation. On a project where the degree of risk is low and health and safety issues are well covered in the principal contractor's policy then a reference to that policy may be sufficient.

*A Guide to Managing Health and Safety in Construction* (HSE, 1995b), suggests a number of specific aspects which should be considered in compiling the construction health and safety plan. These are:

- *Project overview* – this section develops further the information contained in the sub-section of the pre-tender health and safety plan entitled 'nature of the project'.
- *Health and safety standards* – any standards specified by the client, the designer or principal contractor should be included.
- *Management arrangements* – this section should outline the management structure and organization for the health and safety management of the project and specify the key responsibilities of the parties.
- *Contractor information* – this outlines the protocol involved in advising contractors about project risks.
- *Selection procedures* – procedures for assessing the competence and resources of contractors is included in this section.
- *Communication and cooperation* – this section sets out lines of communication and methods for coordinating health and safety.
- *Activities with a risk to health and safety* – hazards identified in the pre-tender health and safety plan are to be communicated to site personnel. Project hazards will have been identified by the principal contractor's risk assessment. This section details how information is to be explained and how identified hazards can be mitigated or reduced. Safe working procedures will be developed for each hazardous activity. These are often contained within a company handbook of generic and specific safe working instructions.

- *Emergency procedures* – this section outlines notification of alarms, escape routes, assembly areas and personnel checks.
- *Accident recording* – this section outlines how the contractor will fulfil his responsibilities under *The Reporting of Injuries, Diseases and Dangerous Occurrences Regulations* (RIDDOR) (HSE, 1995c).
- *Welfare facilities* – the arrangements for all temporary site welfare facilities is included in this section.
- *Training* – this section outlines all health and safety training, including induction training.
- *Site rules* – any site rules and restrictions identified in the pre-tender health and safety plan should be included in the construction phase health and safety plan.
- *Consultation* – this section details the procedures for personnel to raise health and safety issues with their supervisors.
- *Health and safety file* – this details the procedures for the passing of information from the principal contractor to the planning supervisor.
- *Health and safety monitoring* – procedures for inspection and audit are outlined in this section.
- *Project health and safety review* – a report on health and safety and details of any incidents should be outlined in this section.

**Inputs to development**

To facilitate the detailed development of the construction health and safety plan, the principal contractor will require inputs from other project participants. These include the following.

**Contractors**

Contractors can contribute to the construction health and safety plan by providing information concerning:

- identified hazards occurring from their services
- the assessment of risks identified
- the potential measure of control of the identified risks.

Contractors are duty bound to make their personnel aware of all project risks and provide training for those employees where necessary.

**Designers**

Some elements of the design will be finalized only after the construction stage on site has begun. Within the range of the CDM Regulations, the designer continues to have responsibilities for the design irrespective of when the design work is undertaken. Therefore, although the principal contractor assumes responsibility for the construction health and safety plan, and may be implementing it on site, the designer retains responsibility for any design elements still to be passed on to the principal contractor.

Such elements may include, for example, foundations, where there have been variations to those works originally proposed, services, which may have to be changed as a result of unforeseen circumstances, or, finalization of lift installation as a result of detail changes to fabrication design.

### The client

The client will contribute to the development and possibly the implementation of the construction phase health and safety plan. The client will therefore, be included in discussion concerning production matters throughout the construction stage. For example, the principal contractor may request from the client further information on site rules or restrictions, access provision, storage of potentially hazardous materials or substances. Also, the client may be asked to assist with site health and safety training of personnel. Nevertheless, it is the responsibility of the principal contractor to ensure that the construction phase health and safety plan is fully developed and implemented throughout the project.

## Health and safety plan examples

The way in which all the information is compiled into the pre-tender health and safety plan and construction phase health and safety plan can be seen in Appendix II (p. 239) and Appendix III (p. 261) respectively. The examples are in outline form and are presented for illustrative purposes only. The content and levels of detail required when preparing health and safety plans will vary according to the needs of the individual construction project.

## The health and safety plan and management on site

The principal contractor is responsible for planning, managing and controlling health and safety during the construction phase on site. The works should only commence once the construction phase health and safety plan has been fully developed. Furthermore, the plan should always be seen as dynamic and may need to be amended as works proceed to accommodate changing conditions and circumstances or to meet changing requirements in design or specifications.

The principal contractor must use the information in the health and safety plan when managing and controlling the works on site. The key issues within the plan, therefore, need to be translated into the health and safety management system for implementation on site.

The health and safety management system established is centred on creating safe systems of work. These are procedures which are safety and welfare conscious. Their purpose is to ensure the health, safety and welfare of all persons who work on or come into contact with the construction project. Chapter 18 describes the key elements of a principal contractor's health and safety management system.

## Key points

This chapter has identified that:

- The CDM Regulations (HSE, 1994a) require a two-part approach to health and safety planning: (1) the pre-tender health and safety plan; and (2) the construction phase health and safety plan.
- The responsibility for producing the pre-tender health and safety plan rests with the planning supervisor, employed by the client.
- The pre-tender health and safety plan provides information essential to prospective tenderers so that they can plan for and cost necessary safety measures in their project bid.
- The HSE approved code of practice *Managing Construction Health and Safety* (HSE, 1995a) provides a guide to the main aspects for which information can be gathered for producing the pre-tender health and safety plan.
- The responsibility for producing the construction phase health and safety plan rests with the principal contractor.
- The focus of the construction phase health and safety plan is the development of a fully considered and resourced framework for managing the health and safety of all those involved in and interfacing with construction site activities.
- The HSE publication *A Guide to Managing Health and Safety in Construction* (HSE, 1995b) suggests specific aspects which the principal contractor should consider when compiling the construction phase health and safety plan.
- The principal contractor is responsible also for incorporating within the construction phase health and safety plan the consideration of all hazards and risks associated with the activities of any contractors employed on the project.
- The pre-tender health and safety plan and the construction phase health and safety plan come together to provide a detailed record of information for the project – the health and safety file.
- The principal contractor is responsible for managing and controlling health and safety on site and, therefore, should translate the health and safety

plan into a set of management procedures for implementation – a health and safety management system.

- For all hazards identified within the health and safety plan the principal contractor will, within the health and safety mangement system, employ generic or develop specific safe working procedures for implementation by site operatives.

## References

Health and Safety Executive (HSE) (1994a) *The Construction (Design and Management) Regulations 1994*, HMSO, London.

Health and Safety Executive (HSE) (1994b) *CDM Regulations: How the Regulations Affect You*, HMSO, London.

Health and Safety Executive (HSE) (1995a) *Managing Construction for Health and Safety*, HMSO, London.

Health and Safety Executive (HSE) (1995b) *A Guide to Managing Health and Safety in Construction*, HMSO, London.

Health and Safety Executive (HSE) (1995c) *The Reporting of Injuries, Diseases and Dangerous Occurrences Regulations*, HMSO, London.

# 18 System implementation

## Introduction

This chapter presents an outline for the implementation of the principal contractor's health and safety management system. The essence of approach is to assure that there are safe systems of work enacted through implementing safety conscious procedures in key project areas. Seven key elements form the basis of the system which commences with information drawn from the pre-tender health and safety plan and concludes with external information flow to the project health and safety file and internal feedback to the policy making mechanisms of the principal contractor. System records and documentation are essential to effective implementation and therefore suggested pro forma are presented for each system element.

## The principal contractor's health and safety management system

**Safe systems of work**  The health and safety management system established by the principal contractor is centred on creating safe systems of work. Safe systems of work are welfare and safety conscious procedures which are formalized by the implementing organization, understood by all employees, practical in application and, above all, enforceable. The key factors to consider in system implementation are shown in Figure 18.1.

Figure 18.2 illustrates the basic components of a principal contractor's health and safety management system. The system consists of seven key elements. These encompass the principal contractor's duties and responsibilities for project health and safety under the CDM Regulations 1994 (HSE, 1994a) and other specific legislation including: HSWA 1974 (HSE, 1974); MHSWR 1999 (HSE, 1999); PPE Regulations 1992 (HSE, 1992); COSHH 1994 (HSE, 1994b); RIDDOR 1995 (HSE, 1995); and CHSWR 1996 (HSE, 1996).

The system flows from the incoming information from the pre-tender health and safety plan and the construction health and safety plan, through a series of managerial activities to auditing procedures. Feedback mechanisms close the system loop with information directed back to future health and safety

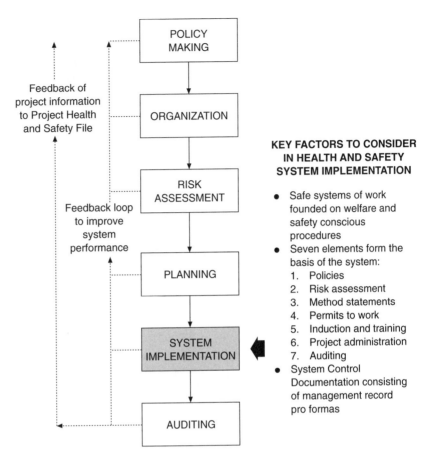

**KEY FACTORS TO CONSIDER IN HEALTH AND SAFETY SYSTEM IMPLEMENTATION**

- Safe systems of work founded on welfare and safety conscious procedures
- Seven elements form the basis of the system:
  1. Policies
  2. Risk assessment
  3. Method statements
  4. Permits to work
  5. Induction and training
  6. Project administration
  7. Auditing
- System Control Documentation consisting of management record pro formas

**Fig. 18.1** Key factors to consider in health and safety system implementation

policy making. An external flow of information passes project information to the health and safety file, a requirement of the CDM Regulations described in Chapter 17.

**Elements of the system**

Seven key elements form the framework of the system, these are:

1. Health and safety management policies
2. Risk assessment
3. Safety method statements
4. Permits to work
5. Safety induction and training
6. Project administration
7. Safety management audits.

### Health and safety management policies

Management policies set out the organization's statement of commitment to health and safety management. They outline how organizational procedures may be established to ensure that safe systems of work are developed and

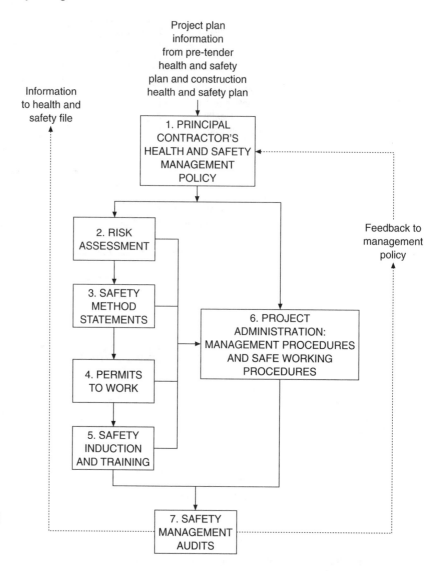

**Fig. 18.2** Principal contractor's health and safety management system

applied to the project. Policy formulation is fundamental to the development of any organization's health and safety management system. It forms the basis for creating the managerial framework, the operational elements within the framework and the assignment of responsibilities to both managers and employees. To establish health and safety management, leadership and the motivation of employees, policies must be clearly stated and communicated. Policies must reflect the organizational commitment at corporate levels and pervade the whole organization through the hierarchy of management to all site employees. These aspects were described in Chapter 14.

To achieve successful health and safety management at project level, the principal contractor must establish a clearly determined set of policies which specify basic procedures and expected minimum standards of performance where possible. Although particular characteristics of health and safety policies will differ among principal contractors and across construction projects, a generic outline can be suggested. A health and safety management system for any construction project should establish a manual of policies and safe working procedures comprising each of the following aspects of the project:

- site access and registration procedures for employees and visitors
- vehicular access to site and associated public safety
- waste management and site egress
- accommodation and welfare facilities
- statutory notes and registers
- hazard and safety signage
- working at heights, in excavations, demolitions
- working with machinery, cranes, hoists, transport
- personal protective equipment
- occupational health (noise, vibration, hazardous substances, pollution, exposure limits to specific conditions, manual handling)
- fire prevention and incidents
- use of electrical supplies, work around services and portable hand tools
- appointment of safety supervisors
- third-party and public safety
- safety induction and training
- safety method statements
- reporting injury, ill-health, near miss incidents and damage.

It is essential that all sets of procedures are carefully controlled. To achieve this a document register can be used. A suggested pro forma is shown in Figure 18.3 (see p. 193).

An example of stated policy for 'waste management and site egress' might be produced as follows:

*Document Section A.1: Waste management*
All contractors and personnel under their control shall be required to:
1. clear building debris and site waste which is the by-product of their operations as work proceeds.
2. maintain a designated and unobstructed route for egress of waste materials.
3. handle debris and waste to designated holding areas as specified in the construction phase health and safety plan.
4. report any circumstances where there is any compromise of 1 and 2 above.

### Risk assessment
It was seen earlier that the designer is not required to devise safe systems of work for the principal contractor. This responsibility is held wholly by the principal contractor. Risk assessment is a fundamental aspect of establishing safe systems of work and is a key element in the principal contractor's health

and safety management system. The function of this system element is to provide management with details of potential hazards and risks and specify when safe systems of work are needed to control them. A suggested pro forma for detailing safe working procedures is shown in Figure 18.4 (see p. 194).

Risk assessment concepts, principles and methodology were described in detail in Chapter 16. Further guidance to risk assessment for safety management is given in the MHSWR 1999 (HSE, 1999). A good basic framework for practical risk assessment on site follows five key steps.

1. Identify the hazards by considering where the works will be undertaken, how the works will be carried out, who will undertake the tasks and what resources and equipment they will be using.
2. Determine who could be in danger from the works and how they could be harmed.
3. Evaluate the potential risk to remove the risk completely or where this is not feasible then consider measures to minimize the degree of risk.
4. Record the findings of risk assessment as an aid to raising awareness and help implement control measures.
5. Review the findings so that lessons can be learned for future situations.

### Safety method statements

Safety method statements provide details of how the safe systems of work are devised around construction operations and tasks. They detail how a job will be carried out, including the control measures which should be applied to minimize potential risk of accident. The statement will help to plan the works and identify the health and safety resources required in their undertaking. A suggested pro forma is shown in Figure 18.5 (see p. 195).

The statements inform the principal contractor's construction project management team of the proposed method of carrying out stage-by-stage tasks. For each task, the health and safety implications can be assessed and precautions taken to ensure a safe working environment. Safety method statements are useful in providing the workforce with essential information on how the works should be carried out in a safe and effective way. They also form the basis for effective monitoring and control mechanisms.

The principal guide to this is the construction health and safety plan outlined in Chapter 17. The function of this system element within the principal contractor's health and safety management system is to provide management with a formalized approach to linking the construction operational sequence to the health and safety dimension. To achieve this it is essential that management responsibilities are clearly determined. A suggested pro forma for detailing this is shown in Figure 18.6 (see p. 196). This consideration is essential both to producing the principal contractor's construction phase health and safety plan and to assessing risk of the proposed work of subcontractors. During the construction phase the safety method statements are an essential reference to guide ongoing site activities. The works of the various contractors need to be effectively coordinated and this is an essential role of management. A suggested pro forma for recording coordinated works is shown in Figure 18.7 (see p. 197).

A safety method statement enables operatives carrying out the work to:

- undertake their tasks in a safe manner
- be aware of the hazards associated with their tasks
- implement control mechanisms to eliminate or reduce the risk of safety incidents.

To achieve these objectives, the safety method statement should:

- describe the operations and tasks that comprise the works
- identify the location of the works
- specify the supervisory arrangements while the works are ongoing
- identify the safety supervisor
- list the plant needs for the operations
- state likely occupational health implications of the work
- specify the precautions to be taken to minimize risk by those carrying out the works.

In developing the safety method statement the required performance of operatives should be considered. In fulfilling the objectives of work tasks, supervisors should inform operatives of expected goals for safe working practice, implement monitoring procedures for assessing performance and provide timely and reliable feedback on works undertaken.

**Permits to work**

A permit to work is a management control procedure which can be issued for almost any construction activity. It is usually applied to activities that give rise to a high risk of danger to those undertaking the work. The issue of a permit to work is a formalized procedure designed to provide a safe system of work, and as such it is issued by a section manager within the principal contractor's organization. Essential aspects of permits to work are that they always relate to tasks where:

- specific training is needed to undertake the task
- the work is considered to be high-risk and denoted as such in the risk assessment records
- the works are complicated in nature or location.

This element of the health and safety management system concentrates on ensuring that the principal contractor has procedures in place to consider, issue, monitor and control permits for hazardous work. A suggested pro forma is shown in Figure 18.8 (see pp. 198–9).

The format of documentation procedures for issuing a permit to work will vary across organizations and even from task to task. However, common aspects to be included are:

- a permit number
- a date and time of issue
- the duration of the permit
- the location of the works
- a description of the hazard, likely risk or potential implications
- precautionary measures to be taken
- any testing or validating procedures of measures imposed
- emergency procedures, signals and reporting
- acknowledgement by the receiving operative in charge of the work
- signing-off of the completed work
- cancellation of the permit by a senior manager (preferably the issuing manager).

### Safety induction and training

The principal contractor should establish management procedures for safety awareness induction for all personnel who are appointed to the project site. This aspect is well covered in the CHSWR (HSE, 1996). Although the mode of delivery will differ among organizations, common elements of safety induction would include:

- specification of the organization's health and safety policy and procedures
- communication of the construction phase health and safety plan and safety method statements relating to particular aspects and characteristics of the works
- specification of expected safety performance standards
- detailing the penalties and disciplinary issues of breaching the procedures
- clarification of any project aspect and encouraging support and coopera-tion among the site personnel.

Induction meetings will integrate the technical aspect of the work with the specified health and safety procedures. They will therefore:

- familiarize personnel with the construction requirements of the works
- inform personnel about the safety aspects of undertaking those works
- determine supervisory management of both the technical and safety aspects.

The MHSWR 1999 (HSE, 1999) require a designated company safety adviser/officer to be present at induction events where the works are complicated and assume high-risk characteristics. At such induction events, a typical agenda will require the safety adviser to:

- remind personnel of the core safety procedures and standards
- identify task/operation-specific safety aspects
- advise on designated access and egress points, traffic management, public welfare and safety, and site boundaries and restricted access
- specify permit to work procedures
- highlight first aid and emergency procedures
- specify safety inspection arrangements and identify safety supervisors present on site
- advise on disciplinary measures for non-compliance with procedures
- specify modes of reporting safety incidents and near incidents
- provide personnel with a safety management manual.

Induction events should be documented and attendance lists compiled. This will reinforce the basic message that safety issues are a serious matter and that the organization is committed to good safety management. Further, it provides a record to follow-up safety training events.

Foremen and gang supervisors are key personnel in the monitoring and control of site activities. They act as essential intermediaries between site management and the workforce. They have a very important role in mentoring, educating and training site personnel and can fulfil this with regular and informative liaison through toolbox talks and on-site works familiarization.

Safety education and training should be seen as an ongoing management procedure with follow-up events being held where, for example, there is a major change to the construction programme, a variation to operational procedures, and new personnel taking-up post on site. A suggested pro forma for recording induction and training activities is shown in Figure 18.9 (see p. 200).

### Project administration: management procedures and safe working procedures

Management procedures set out the methods to be used and the minimal standard requirements of all general activities on the project site. While many aspects of general project administration are common across projects, the CDM Regulations (HSE, 1994a) require that site-specific safe working procedures, or instructions, are established which accommodate the unique requirements of the project. These are often termed 'site rules'.

Management procedures may cover almost any aspect of project administration but, in general, procedures are set in place to cover those aspects of health and safety policy established in the construction phase health and safety plan. These were detailed in the earlier section on health and safety management policies.

Site rules are project-specific and exist to provide safe systems of work where works are to take place in a high-risk environment or where the works themselves are hazardous to those carrying them out. They will cover all hazardous activities identified in the project (see Chapters 16 and 17).

Site rules will be issued in the site health and safety manual applicable to the project to ensure that all personnel are aware of and follow them in association with implementing safe working procedures. It is not an intention within this book to detail all the areas of construction work for which site rules are required. Details are available in authoritative guidance documents produced by the HSE which address the requirements of CHSWR, PPE and COSHH. However, examples of major work aspects will include working at heights, groundworks, working in confined spaces, use of plant and equipment, health hazards such as noise, dust, vibration, use of personal protective equipment, and working with electricity, water and hazardous substances. In each major category there will be sub-categories detailing the site rules for any given situation.

An example of notified site rules for working in a high-risk restricted area might be produced as follows:

*Document Section G.1: Site rules for operating in Zone H (restricted area)*
All personnel must:
1. register upon arrival and departure
2. wear safety helmets and non-spark footwear
3. follow safety procedures and signs at all times
4. be familiar with the use of fire handling equipment and use of breathing apparatus before entering area
5. work only within designated and cordoned areas
6. report any safety matter to Zone H plant safety supervisor.

Safety inspection and reporting are essential aspects of project administration. Safety inspections should be carried out regularly to identify any unsafe working conditions and specify the remedial actions necessary to make them safe. When accidents occur there should be a mechanism in place to report in detail on the circumstances. It is, of course, a requirement under RIDDOR 1995 (HSE, 1995) to report any incident. All incidents should be followed up by a comprehensive investigation and report which examines the causes and presents recommendations to prevent recurrence. Suggested pro forma for these procedures are shown in Figures 18.10, 18.11 and 18.12 (pp. 200–2).

### Safety management audits

An important aspect of the principal contractor's health and safety management system is safety management auditing. An audit is:

- a detailed review of how safety management is being applied
- an analysis of the degree to which the system is being complied with and appraisal to determine the benefits being gained from system implementation
- an identification of systems aspects where a change in procedure can make improvements to overall safety management.

| Document Register | | | | | |
|---|---|---|---|---|---|
| Project: | | | | | |
| Document Reference No. | Title | Issue Date | Amendment Date | Amended Reference | Authorized By |
| | | | | | |
| | | | | | |
| | | | | | |
| | | | | | |
| | | | | | |
| | | | | | |
| | | | | | |
| | | | | | |
| | | | | | |
| | | | | | |
| | | | | | |
| | | | | | |
| | | | | | |
| | | | | | |
| | | | | | |
| | | | | | |
| | | | | | |
| | | | | | |

**Fig. 18.3** Principal contractor's document register: suggested pro forma

| Principal Contractor's Safe Working Procedures |
| --- |
| Project: |
| Document Reference No.: |
| Task or work operation: |
| This Safe Working Procedure has been prepared for the following work |
| Location of work: |
| Description of work: |
| Safe methods to be adopted: |
| Prepared by: |
| Name: |
| Designation: |
| Signature: |

**Fig. 18.4** Principal contractor's safe working procedures: suggested pro forma

| Principal Contractor's Safety Method Statement | | | | |
|---|---|---|---|---|
| Project: | | | | |
| Document Reference No.: | | | | |
| Contractor: | | | | |
| Specialist discipline: | | | | |
| Evaluation of: | | | | |
| The method statement is returned for reconsideration*/accepted*: (*delete as applicable) | | | | |
| Next action: | | | | |
| Assessed by: | | | | |
| Date: | | | | |
| **TEST** | **YES** | **NO** | **IN PART** | **N/A** |
| a. Task/process and area of specialization | | | | |
| b. Sequence of work | | | | |
| c. Supervisory arrangements | | | | |
| d. Monitoring arrangements | | | | |
| e. Schedule of plant | | | | |
| f. Reference to occupational health standards? | | | | |
| g. First aid | | | | |
| h. Schedule for personal protective equipment | | | | |
| i. Schedule of arrangements for demarcation | | | | |
| j. Controls for the safety of third parties | | | | |
| k. Are the assessed high risk or safety critical phases identified with controls specified? | | | | |
| l. Emergency procedures | | | | |

**Fig. 18.5** Principal contractor's safety method statement review: suggested pro forma

| Management Responsibilities | | |
|---|---|---|
| Project: | | |
| Document Reference No.: | | |
| Date issued: | | Page          of |
| Compiled by: | Signed: | Date: |
| Authorized by: | Signed: | Date: |
| Name | Function | Responsibilities |
| | | |
| | | |
| | | |
| | | |
| | | |
| | | |
| | | |
| | | |
| | | |

**Fig. 18.6** Principal contractor's management responsibilities: suggested pro forma

| Activity | Contractor | Start Date | Pre-qualification Questionnaire | | Initial Safety Meeting | Safety Method Statement Evaluation | Risk Assessment | | | | Safety Information given | Authorized to Start |
| | | | Sent | Received and accepted | | | General | COSHH | Noise | Manual Handling | | |
|---|---|---|---|---|---|---|---|---|---|---|---|---|
| | | | | | | | | | | | | |

**Principal Contractor's Site Safety Coordination Record**

Project:

Document Ref No.:

**Fig. 18.7** Principal contractor's site safety coordination record: suggested pro forma

| Principal Contractor's Permit to Work | |
|---|---|
| Project: | |
| Document Reference No.: | |
| Task or Work Operation: | Duration of permit: |
| This Permit to Work is issued for the following: | |
| Is work to be carried out when plant, equipment or systems are in operation? | Yes/No |
| Location of work: | |
| Description of work (specific hazards): | |
| Precautions to be taken: | |
| Extra precautions to be taken if plant and equipment is being used: | |
| Additional permits:<br>● Hot work<br>● Electrical<br>● Confined space<br>● Other | |
| **Authorization** | |
| Name of person issuing Permit: | |
| Designation: | |
| Signature: | |
| Time: | Date: |

**Fig. 18.8** Principal contractor's permit to work: suggested pro forma

| Receipt |
|---|
| Name: |
| Designation: |
| Signature: |
| Company: |
| **Clearance** |
| The work stated above has/has not been completed.<br>Details if not completed: |
| Name: |
| Designation: |
| Signature: |
| Company: |
| **Cancellation** |
| Permit to work is cancelled |
| Name: |
| Designation: |
| Signature: |

| Date: | Time: |
|---|---|

**Fig. 18.8** (*cont'd*)

| Principal Contractor's Induction and Training | | |
|---|---|---|
| Project: | | |
| Document Reference No.: | | |
| Contractor: | | |
| Type of training: | | |
| Name of trainee | Trainer | Date of training |
| | | |
| | | |
| | | |
| | | |
| | | |
| | | |
| | | |
| | | |
| | | |
| | | |
| | | |
| | | |
| | | |

**Fig. 18.9** Principal contractor's induction and training: suggested pro forma

| Principal Contractor's Site Safety Inspection | | | |
|---|---|---|---|
| Project: | | | |
| Document Reference No.: | | Date: | Time: |
| Location: | | | |
| Any unsafe conditions or work: | | | |
| Remedial action: | | | |
| Further action to be taken | By (Named person) | Date | Complete |
| | | | |
| Inspected by (Safety Inspector):<br><br>Action authorized by: | | | Date: |

**Fig. 18.10** Principal contractor's site safety inspection: suggested pro forma

**Principal Contractor's Accident Report**

Project:

Document Reference No.:

Page                    of

| Injured person | Accident | Person reporting accident |
|---|---|---|
| Name: | Date:                    Time: | Name: |
| Home address: | Location: | Home address: |
| | Work process involved: | |
| | | |
| | Cause (if known) | |
| Occupation: | Details of injury: | Occupation: |
| | | Signature: |
| | | Date of report: |
| Name: | Date:                    Time: | Name: |
| Home address: | Location: | Home address: |
| | Work process involved: | |
| | | |
| | Cause (if known): | |
| Occupation: | Details of injury: | Occupation: |
| | | Signature: |
| | | Date of report: |

**Fig. 18.11** Principal contractor's accident report: suggested pro forma

| Principal Contractor's Incident Investigation Report | | |
|---|---|---|
| Project: | | |
| Document Reference No.: | | |
| Parties involved: | | |
| Location of incident: | | |
| Date of incident: | | Time of incident: AM/PM |
| Type of incident: | | |
| Potential severity: | ☐ Major ☐ Serious | ☐ Minor |
| Probability of recurrence: | ☐ High ☐ Medium | ☐ Low |
| Description of how incident occurred: | | |
| Immediate causes: what unsafe acts or conditions caused the event? | | |
| Secondary causes: what human, organizational or job factors caused the event? | | |
| Remedial actions: recommendations to prevent recurrence: | | |
| Signature of investigator: | | Date: |
| Follow up action/review of recommendations and progress | | |
| | | |
| | | |
| | | |
| | | |
| | | |
| Name of reviewer: | | |
| Position/title of reviewer: | | |
| Signature of reviewer: | | Date: |

**Fig. 18.12** Principal contractor's safety incident investigation: suggested pro forma

A safety management audit may be conducted in-house in association with the organization's corporate safety team or could be conducted by specialist consultants hired for the task. The approach of the audit should be to:

- evaluate the current safety policies
- approve the safe working procedures currently specified
- appraise the construction phase health and safety plan
- identify the control mechanisms in place to ensure compliance with project requirements (identified above)
- review the current construction programme, site structure and organization and management personnel responsible for safety aspects
- appraise safety management reports of incidents and near incidents
- provide a detailed report on the findings of the audit.

Further details on safety auditing are presented in Chapter 19.

## Key points

This chapter has identified that:

- A principal contractor's health and safety management system is based on creating safe systems of work, i.e., procedures which are welfare and safety conscious.
- Seven key elements form the basis of a system, these are: (1) policies; (2) risk assessment; (3) method statements; (4) permits to work; (5) induction and training; (6) project administration; (7) auditing.
- Specific requirements for health and management for the project will be determined by the information derived from the health and safety plan.
- The recording of information is essential to system implementation and therefore simple yet comprehensive pro forma documentation should be utilized by management on site.

## References

Health and Safety Executive (HSE) (1974) *The Health and Safety at Work, etc. Act 1974*, HMSO, London.

Health and Safety Executive (HSE) (1992) *The Personal Protective Equipment at Work Regulations 1992*, HMSO, London.

Health and Safety Executive (HSE) (1994a) *The Construction (Design and Management) Regulations 1994*, HMSO, London.

Health and Safety Executive (HSE) (1994b) *The Control of Substances Hazardous to Health Regulations 1994*, HMSO, London.

Health and Safety Executive (HSE) (1995) *The Reporting of Injuries, Diseases and Dangerous Occurrences Regulations 1995*, HMSO, London.

Health and Safety Executive (HSE) (1996) *Construction (Health, Safety and Welfare) Regulations 1996*, HMSO, London.

Health and Safety Executive (HSE) (1999) *The Management of Health and Safety at Work Regulations 1999*, HMSO, London.

# 19 Auditing

## Introduction

The auditing and review of the health and safety management subsystem is essential to maintaining the effectiveness of both the corporate organization and the control procedures utilized within the project organization. Independent auditing provides information to the organization which can be linked back to company policy making so providing feedback which completes the loop in the health and safety management subsystem. Auditing focuses on the dynamics of the health and safety management approach and the behavioural aspects of the people who implement the approach. This chapter presents an insight into the key elements of auditing by the organization.

## Safety management auditing

An important aspect of the principal contractor's health and safety management approach is safety management auditing. Auditing is the feedback loop within the safety management subsystem that enables the organization to assess the effectiveness of its implementation and consider, if necessary, its further development to effect improvement (see Figure 19.1).

A safety management audit is:

- a detailed review of how safety management is being perceived, understood and applied within the project organization
- an analysis of the degree to which the management subsystem is being complied with by project practices
- an appraisal to determine the systems benefits being gained from that subsystem implementation
- the identification and analysis of those elements of the health and safety subsystem where changes in procedure and/or behaviour can offer improvements to safety management approach.

Safety management auditing is essential to maintaining the organization's health and safety control procedures on site. All control mechanisms can tend

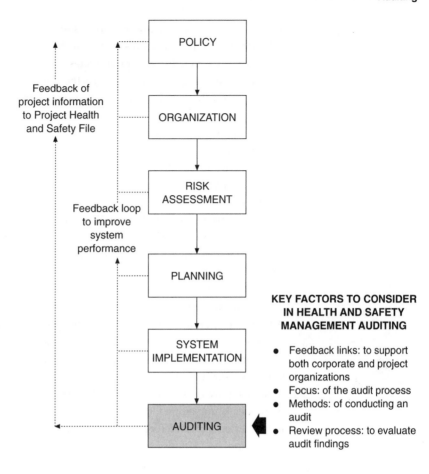

**Fig. 19.1** Key factors to consider in implementing the auditing process

to become less efficient over an extended period of time as application becomes complacent or outmoded. Control mechanisms, like other elements within management subsystems, need to be checked regularly to ensure that they maintain their currency and thereby their effectiveness. There are two key elements to safety management auditing, the audit process itself and the process of review.

## Auditing

An audit aims to provide an independent and therefore an unbiased and balanced assessment of the effectiveness of the health and safety management subsystem in implementation. Independent auditing augments the subsystem monitoring mechanisms by providing information to the organization which can be linked back to company policy making, organization, risk assessment and planning. In so doing the organization has the opportunity to consider and review its entire subsystem to ensure its implementation is consistent with other organizational subsystems, such as quality or environmental procedures used across different project sites and the policy and objectives of the corporate organization which supports the company's core business. An audit may focus on the management mechanisms of the system's implementation or behavioural characteristics of the managers and workforce that enact the system's procedures and working instructions (see Chapter 13).

An audit may gather information on health and safety performance in both quantitative and qualitative forms. Some organizations seek to generate audit information based on numerical data so that performance can be compared annually. Others gather anecdotal information based on experiences and lessons learnt in implementing health and safety management. The important issue in information gathering is that it must allow sufficient reliable information to be available to management which gives a comprehensive picture of how effectively the health and safety subsystem is at handling project risks.

A safety management audit may be conducted in-house, that is by the project organization in association with the corporate health and safety adviser. Alternatively, an audit may be conducted by specialist consultants hired for the task. The methods used can be a proprietary approach which implements a standard system or the organization's self-developed approach, or even a hybrid version based on both methods. The essential issue is that it should be undertaken by competent and informed persons who are independent of the work area being audited. It should also be remembered that each organization is individual and therefore there is no one right approach to auditing which can be applied to all organizations.

**Focus of audits**    A general approach to health and safety management will focus on policy, organization, planning, and review mechanisms. This will involve the evaluation of the:

- current health and safety policies
- responsibilities of health and safety advisers and supervisors and first-line construction supervisors
- safe working procedures
- construction health and safety plan
- control mechanisms to ensure compliance with the above
- construction programme, project structure and organization and personnel deployment
- safety management reports of investigation of incidents and near incidents
- review mechanisms and information forwarding to future projects.

**Audit methods**    While an organization will adopt its own approach to health and safety management auditing, there are broadly two general methods:

1. Longitudinal analysis
2. Cross-sectional analysis.

These are complementary methods of gathering information from the health and safety subsystem by looking at specific elements in different ways. For example, a longitudinal approach would involve an in-depth examination of one specific element of the health and safety management subsystem, such as 'planning' or 'organization'. The whole process of operation of the element in question would be the focus of the audit and a track followed throughout its operation within the organization. A cross-sectional approach would involve the in-depth examination of a particular activity within an element, for example, looking at the remit of the safety adviser within the planning processes.

## Review

The process of review is concerned with evaluating the findings of an audit, together with any interim inspections of the health and safety management subsystem, to determine its performance in application. An essential aspect of review is the identification of elements within the management subsystem which require remedial action and determining how and by whom such action is to be implemented.

Generally, the aim of the review process is to allow management to reflect upon organizational policies, organization and planning in the light of the most recent organizational performance. Review therefore should ensure:

- the development of health and safety policies
- the development of an effective organization to support the pursuit of existing and future policies
- the development of performance standards, inspection and reporting mechanisms to support the assessment of actions implemented in pursuit of policies.

The review of health and safety performance should be based on factual information obtained from both reactive and active monitoring of site activities. Review is not a static concept but rather a dynamic one in which monitoring is a continuous process. Review is also not an annual activity but one which takes place periodically and consistently throughout the calendar year and contributes to detailed annual review. Reviews may be conducted monthly, three monthly, half yearly and annually. An organization must decide for itself the frequency of review at both corporate and project level.

Any review will examine the information gathered by the auditing process. It is important therefore that the organization develops its own performance indicators for which it gathers information during auditing and against which performance is assessed during review. A number of performance indicators are key to health and safety management review.

- The degree of compliance with set performance standards.
- The identification of activities where performance standards are inadequate.
- The degree of fulfilment of particular objectives, for example those specific to an individual project.
- The number, frequency, type, cause and impact of accident and welfare related incidents.

The review of this type of information is essential to providing detailed feedback through the health and safety management subsystem to those in the organization responsible for policy making, organization, risk assessment, planning, and implementation. Effective review within a successful health and safety management subsystem will emphasize the importance of constantly measuring against set performance indicators and reviewing accordingly. A measure of the success of audit and review procedures is the degree to which they are naturally incorporated into everyday duties rather than being managerially imposed.

Review may also seek to benchmark the organization's performance against other organizations. This might include other contracting organizations or even organizations in other industry sectors. This is becoming more commonplace and it is not uncommon to see health and safety performance review featuring in the publicity materials and annual reports of many construction organizations.

## Key points

This chapter has identified that:

- The process of auditing is essential to effective health and safety management as it enables the organization to assess the effectiveness of its implementation and effect improvements where necessary.
- Any management control mechanism can become less efficient over time and auditing allows periodic checks to ensure they maintain their currency.
- Auditing aims to provide an independent and unbiased assessment of the effectiveness of management implementation to augment the subsystem monitoring mechanisms.
- The audit process may gather information on health and safety performance in quantitative and qualitative forms such that performance can be compared from one period of time to another.

- The process of review evaluates the findings of a health and safety management audit, identifying elements requiring remedial action and determining necessary action.
- Review is a dynamic mechanism in which monitoring of the health and safety management subsystem is a continuous process providing reliable and consistent feedback into company policy making, organization, risk assessment, planning and management implementation.

## Introduction

Specialization is intrinsic to the total construction process. Large and complex projects are characterized by teams of specialists, which traditionally include planners, estimators, surveyors, engineers and construction managers. In recent times these specialists have been joined by an additional host of managers, the most prominent being responsible for health and safety, quality, and environmental impact. Today's construction projects could not operate without such specialization. Each specialist contributes to the project through implementing particular management procedures, or systems, or by following established professional working practices. Specialists guide the various stages of the total construction process and the systems which they adopt assist them to maintain the many, varied and complex arrangements that need to be made (see Chapter 11). Moreover, clients and corporate management of organizations require an assurance that organizational and project procedures are being followed and this is particularly pertinent where independent auditing mechanisms assess performance.

Safety, quality and environment are aspects of construction which utilize management subsystems that need to be maintained and all are subject to increasingly stringent regulation, monitoring and assessment. However, as more systems are introduced to the construction process to meet ever demanding requirements, the boundaries between the systems can become indistinct and control procedures can become vague, in particular at the project level during the production phase, (see Chapter 12). At that stage many of the intended benefits of the individual management systems may be lost. However, the complexity of multiple management systems could be replaced by a more straightforward integrated management system. Such a system may help to dispel the problems experienced at the interface between specialists while providing all the support services necessary to sustain project performance.

Chapters 11 and 12 introduced the concept of the parent management system, within which various subsystems can be structured and organized.

Indeed, some larger contracting organizations have already taken the route to using a certificated quality management system as the host system for developing a formalized health and safety management approach. In 1992, the European Construction Institute first published *Total Project Management of Construction Safety, Health and Environment* (ECI, 1992), or the SHE (Safety, Health and Environment) approach. This is a prominent example of total management system development. In addition, the British Standards Institution (BSI) is developing a mechanism for Integrated Management System Assessment (IMSA), in which management systems may be certificated in any combination of quality, environment and safety. This chapter considers the potential for contracting organizations to develop integrated management systems.

## Management specialization

While specialization is essential to the construction processes it can create problems (Arnold, 1994). On a smaller project, coordination of specialist activities is relatively simple, but, on a larger project with a greater number of specialists coordination can be difficult. It becomes increasingly so as more and more specialists are involved. Appreciating the interface between specialist disciplines is essential to avoid ambiguity or confusion (Griffith and Sidwell, 1995). Equally, it is important to identify any overlap between disciplines as this may point the way forward for restructuring the various inputs to create an integrated management system.

## Management support services

Safety, quality, and environment are three key areas in construction project management, but, they must be seen in context. Each is a support service, i.e. a service which can be paramount to project team effectiveness but which is not in itself producing direct financial income (see Chapter 11). The importance of such services should not be underestimated as they fulfil essential functions and they interact with and influence the other functions (Griffith, 1992). However, their role is often seen in isolation and this is to miss the point that they fundamentally exist to provide an efficient and effective service to the project. Their position as support services does mean that management should look at the possibilities for reducing operational costs while maintaining a high level of delivery and for this reason an integrated system, rather than multiple and separate systems, might be preferred.

## Quality management systems

Total quality management (TQM) is recognized as 'having a claim to be one of the very oldest recorded management concepts' (McGeorge *et al.*, 1996). Quality has traditionally been interpreted as 'ability to satisfy needs' (BSI, 1971), 'conformance to requirements' (BRE, 1978), and 'fitness for purpose' (CIRIA, 1985), although recent trends have seen a more holistic understanding of quality emerging in terms of providing customer satisfaction (Griffith, 1990; Ross, 1993). This orientation towards the customer has focused the attention of quality management as a process which links to the various stages of the total construction process and which underpins all activities and business of an organization involved in any of those stages (Armstrong, 1993).

The development of formal quality management systems (QMS) has evolved from the need to comply with worldwide quality standards, e.g. ISO 9000 (ISO, 1987). Compliance with such standards implies that an organization follows documented procedures and working practices which are subject to formal measurement, audit and review, this being conducted internally and by an independent third party. Such a management system imposes rigour to the core elements of structure, organization and management for quality within the construction processes. It is these same elements that present the basis for safety management and environmental management.

## Safety management systems

Current thinking in the management of safety is following the trend set in quality management as discussed previously. Guidance literature published by the HSE advocates a systems approach to safety management as presented in this book. Such a system can be developed to meet BS 8800 (BSI, 1998), the UK specification for health and safety management systems (H&SMS). The HSE defines an accident as 'any unplanned event that resulted in injury or ill health of people, or damage or loss to property, plant, materials or the environment or a loss of business opportunity'. This interpretation clearly highlights the 'unplanned event' and the assessment of risk as key elements in the management of safety and one which is also common to the management of environmental impact and quality.

While good construction management practices have, in the main, addressed the requirements for safe working within construction, a formalized systems approach to safety management has recently begun to emerge. In the UK this has followed the introduction of the Construction (Design and Management) Regulations 1994 (HSE, 1994) (see Chapter 1). Meeting these Regulations is contingent upon the contractual parties implementing a 'safety plan' and maintaining a 'safety file' for their project. To aid the fulfilment of these requirements a systems approach is suggested. This can provide the necessary structure and rigour to safety planning, procedures and management both within the corporate organization and in the project situation as suggested throughout this book.

## Environmental management systems

ISO 14001 is the international standard for environmental management systems (EMS) (ISO, 1994). This standard specifies the basic requirements for the formulation, development, implementation and maintenance of a management system directed towards compliance with an organization's stated environmental policy and objectives and in meeting current environmental legislation. Until the introduction of ISO 14001, formalized applications of environmental management within the construction industry had been limited to a relatively small number of projects with high contract values and harbouring considerable environmental risk. Support for EMS is growing as awareness within the construction industry increases. In the same way that compliance with ISO 9000 is an accepted 'hallmark' of an organization's commitment to quality of service or product, so an EMS meeting the specification of ISO 14001 has become the accepted benchmark for environmental management.

The environmental standard encourages the development and implementation of an EMS based upon existing management systems, for example a quality management system (QMS). Research by Griffith (1995) identifies that contracting organizations are seeking to develop environmental management systems based directly on their certified quality management systems. In fact, this is encouraged by the applicable standards. While meeting the distinct requirements of the separate standards the current focus is on integrating the core elements of the systems to minimize duplication of development effort, assist implementation and reduce operational costs.

## Purpose, structure and characteristics of quality, safety and environmental systems

It is not intended within the scope of this chapter to digress into the detailed concepts and applications of systems theory (Argyris, 1960), nor to become involved in philosophical discussions relating the systems approach to various contemporary management concepts, such as those presented by Kelly and Male (1993) and Chen and McGeorge (1994). However, it is contended that the management of safety, quality, and environment are linked by a common base in systems theory (Griffith, 1994) and that strong philosophical and structural linkages exist.

**Purpose**   The principal purpose of each of the management systems is to provide an open system – one which appreciates and interacts with its environment – to plan, monitor and control its respective sphere of interest within the organization (Lavender, 1996). For example, environmental management is concerned with the policy, strategy and procedures that form the organization's response to its environmental situation in the course of running its business. Similarly, the direct link to organizational ethos, culture and approach which is demonstrated in environmental management is shared by both quality and safety management. Each advocates a holistic or 'whole organization' philosophy, presents a well-defined framework that structures organizational commitment and resources, and is formal in its approach to developing uniform directives and operational procedures. Each is proactive in nature, seeking to be preventive, or quickly reactive, rather than being retrospective in its actions.

These contemporary yet relatively modern management support services are rapidly becoming paramount to the competitive position of contracting organizations. Many organizations are greatly aware of the need to embrace these aspects, if not to secure competitive advantage then certainly to avoid competitive disadvantage in the marketplace. Pre-qualifying for project quality, safety record and environmental safeguard within the tendering process is increasing in popularity in many countries (Griffith, 1995). These concepts are therefore becoming critical to any principal contracting organization's core business thinking.

**Structure**   Each management aspect is well structured by recognized and authoritative national and international standards. Quality management is embraced by ISO 9000 (ISO, 1987), environmental management by ISO 14001 (ISO, 1994) and safety management is currently encompassed within BS 8800 (BSI, 1998). These standards, within their respective spheres of activity, specify the basic requirements for the formulation, development, implementation and maintenance of a structured management approach, or system, directed towards compliance with an organization's policy and objectives and in meeting current applicable legislation.

A further aspect of common structure is that all three management spheres, under the influence of the respective standards, follow documented procedures and working practices. These range from guiding the business operation of the whole organization to the specifics of project procedure or service delivery, in the context of a contracting organization this would be the project site. However, the impact of these management systems are wider than merely providing a response to any standard or Regulation, in fact they can have a positive effect upon the perpetuation of the business as a whole, (see Chapter 11).

**Characteristics**   In combination with the common aspects of structure, the characteristics of safety, quality and environmental management relate strongly to the systems application propounded by Armstrong (1993). He identifies the discipline within an organization of defining objectives, the establishment of measurable indicators of success in meeting objectives and the development of procedures to facilitate the pursuit of objectives. Exposure to the broader internal and external environments determine that audit and review are also prominent. Such characteristics are reflected in the need to manage change and its associated risks in a dynamic way. Safety, quality, and environmental management systems help organizations come to terms with changing market demands and the assessment of business risk. Important characteristics of organizational development follow from these systems, for example, corporate focus, credibility in the marketplace, reduced liability, and organizational efficiency and effectiveness (see Chapters 14 and 15). So, both the management system and the organization should help shape each other in a symbiotic way.

## Compatibility of the systems

Specialization within the corporate construction organization and construction project management is traditional. Safety, quality, and environmental management, like all forms of core and support services, have developed individually as the need for more structured and rigorous management procedures has increased. In addition, development has been a function of individual areas being championed by both industry as a whole, principally through focused client demands, and by organizations within the industry, led by contractors, as increasingly stringent standards and Regulations have evolved.

Although certain aspects and tasks of quality, safety and environmental management are quite specific, many aspects are common to any management systems approach. ISO 9000, ISO 14001 and BS 8800 share key elements in structure and advocate procedures which encourage compatibility, while recognizing and accommodating their individual and dedicated interests. Key elements which feature in the standards and should be reflected in any management system are policy, aims and objectives, programmes, documentation, working procedures, record keeping and audit and review. These key elements are sufficiently compatible to form the basis of an integrated management system at corporate level which can be extended to construction project specific applications.

## Synergistic links

It is suggested that there is a strong synergistic link between safety, quality, and environmental management systems which results not only from the strong philosophical similarity and common elements of structure but also because each system is intended to be dynamic in nature. Each evolves through a process of planning, monitoring, controlling and review when applied to the project situation. This dynamicism develops intrinsically from an effective systems approach, one which is actively supported by corporate management and well understood and accepted by project teams. The key inputs of active leadership, the commitment to specific policy and goals, the clear definition of roles and procedures and frequent review and development combine to reinforce the synergistic link.

The real synergy of an integrated management system, because it is also true of the individual systems, is the recognition of the key characteristic which can be used as the focus for system development and application. This characteristic, which was identified earlier, is the 'unplanned event'. This is common to quality, safety and environmental impact. The unplanned event is paramount as it highlights the principal element upon which to structure the integrated management system – assessment of risk. If the assessment of risk is the core of the system, a vehicle must be identified for facilitating risk assessment. Within construction management this vehicle is invariably 'the planning process' – appreciated in the broadest sense as planning, monitoring, control and review. Assessing risk in relation to quality, safety and environmental management is an essential task throughout the planning process (see Chapters 16 and 17).

## Towards an integrated management system

**Awareness**  Traditionally, there has been a need for separate and specialist quality, safety and, more recently, environmental management sections, or departments, within organizations. This is because each management function has evolved at different times and it has been difficult for both corporate and project management to rapidly understand all the concepts involved. A major problem with this situation is that each management area has developed on its own and has become, perhaps, too introspective, concentrating on its own performance in application rather than focusing on 'the service' provided to the core business. Moreover, problems are compounded because the corporate organization assumes that dedicated departments take complete responsibility and therefore they themselves are not as participative as they should be in getting the best out of the support services.

In addition, the three management systems are relatively new and it is likely that some managers would experience difficulty in getting to grips and coming to terms with new management demands and methods. Some might prefer to dwell in the past with, what has been up until now, quite acceptable management styles and methods. The point is that, today, these systems are 'additional' to the good management practices of the past and are necessary to meet a much wider acceptance of total construction performance.

However, maintaining individual and separate management systems is time consuming and costly and this is accentuated if the systems being supported are not doing their job completely efficiently and effectively. As organizations have to provide these multiple support services and managers, predominantly in the construction project situation, must use these services, the question is: can the individual management systems for quality, safety and environmental impact be drawn together to form an integrated system? If, for example, quality, safety and environmental management had evolved at a similar time it is likely that the concepts, principles and procedures would have been the product of synchronized development. Because they developed at different times they are disparate and integration is now being made in bursts as one new concept attempts to catch up with the one that came before. This is not in itself inappropriate for each new system can develop with the experience of application rather than rely upon conjecture and anticipation.

**System development and outline procedures**

Contracting organizations do not have to develop and implement extensive and complex management systems. This is true even though it is intended to encompass multiple areas of management within one system. It is acceptable, and indeed far more practical, for an organization to build up the system by developing new, or amending existing, procedures and bringing these within the scope of the integrated system over time and in steps. Although, for the reasons stated previously, it is advisable to follow the system specification of a recognized standard, by no means is a standard system necessary. It is perfectly feasible to develop a bespoke in-house system for environmental management based on a quality system (CIRIA, 1995).

There are a number of prerequisites to integrated systems development. These apply at corporate level and at project level and focus upon management commitment and management planning respectively. Also, while the integrated system implies generic procedures, a clear focus on the specifics of quality, safety and environmental impact must be retained. Essentially, the integrated system will manage each in a dedicated way but in systems terms be brought together under one administrative umbrella.

**Corporate level**

One of the principal impediments to organizations delivering such initiatives quickly and successfully is that they will, at first, see an integrated management system as providing soft benefits rather than hard benefits. As a consequence of this, direction and commitment can be lacking. It will always be difficult to quantify potential benefit in financial terms and this is exacerbated as such systems really exist to assess risk and implement procedures which seek to prevent the unplanned event occurring rather than mitigating the effects of an event, where at least the expenditure can be equated to actions taken.

Development vision and support for the system must be demonstrated by corporate management and this needs to permeate the entire organization. To ensure that the system has an optimum chance of being effective there needs to be:

- demonstrable commitment from executive corporate management
- a clear statement of organizational policy and this should be circulated throughout the organization
- employee ownership of the system through involvement in development and implementation
- identified goals against which performance can be compared
- adequate resources to facilitate the system framework and operational management
- ongoing review and improvement to system application.

**Project level**  An organization's corporate approach to any management system development is only as good as its successful implementation to its projects. A structured approach towards the management of quality, safety and environmental impact will better facilitate site management. An integrated system should aid this through specific steps being taken in three clearly defined stages. These were identified in a study of environmental management undertaken by Griffith (1994) and can be applied to the integrated system. The stages, and the steps to be taken, are as follows.

### Project appraisal

This is concerned with the pre-construction stage where risk assessment and planning are prominent. The following steps should be undertaken.

- A quality, safety and environmental impact risk assessment as part of the tendering process.
- The identification of key quality, safety and environmental impact issues that need to be addressed on site.
- The development of quality, safety and environmental impact plans.
- The distribution of good practice guidelines to staff relevant to the key issues identified.
- The determination of audit procedures between the corporate organization and the project site.

### Project familiarization

This is concerned with commencement on site, where the understanding of all project team members is paramount. The following steps should be undertaken.

- The briefing of all project staff in the quality, safety and environmental impact issues identified in project appraisal.
- A site tour to familiarize the construction team with the project and its relationship to the risks in all areas.
- Training in the use of the procedures that will be used to plan, monitor and control project quality, safety and environmental impact.

### Project (production) management

This is concerned with control during production, where communication of system procedures is foremost. The following steps should be implemented.

- Practice notes on good quality, safety and environmental management.
- An item checklist for quality, safety and environmental risk assessment and risk monitoring.
- Self-audit/review sheets for system managers.
- Guidance notes on potential actions should issues or problems arise.
- References to higher management (corporate level) who can be called upon to respond quickly to issues and problems should the need arise.

Simple site reporting mechanisms should be adopted as pro forma checklists to assist the production management stage. These can formally record the following:

- occurrence of any deficiency in quality, safety or environmental impact
- location of the incident
- reason for the occurrence
- any action taken
- review of action to assess effectiveness
- further action needed
- notifiable nature of specific incidents, for example the report to authorities of accidents to persons or breaches of environmental regulation.

An integrated management system for project safety, quality and environment briefly outlined can be given sufficient structure and rigour in application to lay the general foundation for certificated management systems. Specific aspects of the system will need to be addressed in line with the requirements of particular standards and current legislation. An organization that does not currently have a system for managing its quality, safety and environmental matters would do well to consider adopting such a general approach from which it can build up a set of systems which could, in time, be put forward for certificated status in the various areas of management.

On an everyday basis, construction managers have to manage project quality, safety and environmental impact concurrently. In the past, while specialization, to meet the evolution of particular management concepts, has led to separate management systems being developed, it is evident that there could be benefits to be accrued from integrating these independent systems into one management system which would be more easily understood and more simply administered (Arnold, 1994). In the future it is quite possible that the various systems used by a construction management will simply be referred to as a 'total management system' and that this system will embrace all the key areas of construction project management.

Principal contracting organizations may find that they are able to develop elements of a management system almost on a piecemeal basis and build it up over time as experience grows and their needs expand. Some organizations may already have procedures for good practice in place and simply need to develop the structured framework to turn them into a recognized and integrated management system (CIRIA, 1995).

A vehicle for implementing an integrated system is undoubtedly the planning process which, as a main principle, involves risk assessment as a focus when developing the project plan which encompasses quality, safety and environmental impact. The plan, in broad terms, should focus on the unplanned event with workmanship and zero-defect performance within quality management, hazards within safety management, and threats to the project's surroundings within environmental management.

Above all, it is essential that any system is translated into simple and workable procedures and the development steps referred to in this book would go some way towards achieving this. Principal contracting organizations would likely become more efficient and effective if an integrated management system was to be adopted in the future. This would not leave managers immersed in the current situation of meeting the requirements of many disparate organizational systems which they may not fully understand and, therefore, may not manage effectively.

## References

Argyris, C. (1960) *Understanding Organizational Behaviour*, Dorsey, New York.

Armstrong, M. (1993) *A Handbook of Management Techniques*, Nichols, New Jersey.

Arnold, R. (1994) *Management of Specialists and the Role of Integrated Safety and Quality Systems*, Croner's Management of Safety, Issue No. 6, Croner Publications, London.

Building Research Establishment (BRE) (1978) *A Survey of Quality and Value in Building*, BRE, Watford.

British Standards Institution (BSI) (1971) *BS 4778: Specification for Quality*, BSI, London.

British Standards Institution (BSI) (1998) *BS 8800: Specification for Health and Safety Management Systems*, BSI, London.

Chen, S. E. and McGeorge, D. (1994) 'A systems approach to managing buildability.' *Australian Institute of Building Papers* (5), pp. 76–86.

Construction Industry Research and Information Association (CIRIA) (1985) *Quality Assurance in Civil Engineering*, CIRIA, London.

Construction Industry Research and Information Association (CIRIA) (1995) *Effective Environmental Management*, CIRIA, London.

European Construction Institute (ECI) (1992) *Total Project Management of Construction Safety, Health and Environment*, Thomas Telford, London.

Griffith, A. (1990) *Quality Assurance in Building*, Macmillan, Basingstoke.

Griffith, A. (1992) *Small Building Works Management*, Macmillan, Basingstoke.

Griffith, A. (1994) *Environmental Management in Construction*, Macmillan, Basingstoke.

Griffith, A. (1995) 'The current status of environmental management systems in construction.' *Engineering, Construction and Architectural Management* (2), No. 1, pp. 5–16.

Griffith, A. and Sidwell, A. C. (1995) *Constructability in Building and Engineering Projects*, Macmillan, Basingstoke.

Health and Safety Executive (HSE) (1994) *The Construction (Design and Management) Regulations 1994*, HMSO, London.

International Standards Organization (ISO) (1987) *ISO 9000: Specification for Quality Systems*, HMSO, London.

International Standards Organization (ISO) (1994) *ISO 14001: Specification for Environmental Management Systems*, HMSO, London.

Kelly, J. and Male, S. P. (1993) *Value Management in Design and Construction – The Economic Management of Projects*, Spon, London.

Lavender, S. (1996) *Management for the Construction Industry*, Longman/CIOB, Harlow.

McGeorge, W. D., Chen, S. E., Barlow, K., Sidwell, A. C. and Francis, V. (1996) 'Current management concepts in the construction industry – where to from here?' *Australian Institute of Building Papers* (7), pp. 3–12.

Ross, J. E. (1993) *Total Quality Management – Text, Cases and Readings*, St Lucie Press, Florida.

# Appendices

**SECTION 1**
**GENERAL PROVISIONS**

*Article 1*
*Object*

1   The object of this Directive is to introduce measures to encourage improvements in the safety and health of workers at work.

2   To that end it contains general principles concerning the prevention of occupational risks, the protection of safety and health, the elimination of risk and accident factors, the informing, consultation, balanced participation in accordance with national laws and/or practices and training of workers and their representatives, as well as general guidelines for the implementation of the said principles.

3   This Directive shall be without prejudice to existing or future national and Community provisions which are more favourable to protection of the safety and health of workers at work.

*Article 2*
*Scope*

1   This Directive shall apply to all sectors of activity, both public and private (industrial, agricultural, commercial, administrative, service, educational, cultural, leisure, etc.).

2   This Directive shall not be applicable where characteristics peculiar to certain specific public service activities, such as the armed forces or the police, or to certain specific activities in the civil protection services inevitably conflict with it.

   In that event, the safety and health of workers must be ensured as far as possible in the light of the objectives of this Directive.

*Article 3*
*Definitions*

For the purposes of this Directive, the following terms shall have the following meanings:

(a) worker: any person employed by an employer, including trainee and apprentices but excluding domestic servants;

(b) employer: any natural or legal person who has an employment relationship with the worker and has responsibility for the undertaking and/or establishment;

(c) workers' representative with specific responsibility for the safety and health of workers: any person elected, chosen or designated in accordance with national laws and/or practices to represent workers where problems arise relating to the safety and health protection of workers at work;

(d) prevention: all the steps or measures taken or planned at all stages of work in the undertaking to prevent or reduce occupational risks.

*Article 4*

1   Member States shall take the necessary steps to ensure that employers, workers and workers' representatives are subject to the legal provisions necessary for the implementation of this Directive.

2   In particular, Member States shall ensure adequate control and supervision.

## SECTION II
## EMPLOYERS' OBLIGATIONS

*Article 5*
*General Provision*

1   The employer shall have a duty to ensure the safety and health of workers in every aspect related to the work.

2   Where, pursuant to Article 7(3), an employer enlists competent external services or persons, this shall not discharge him from his responsibilities in this area.

3   The workers' obligations in the field of safety and health at work shall not affect the principle of the responsibility of the employer.

4   This Directive shall not restrict the option of Member States to provide for the exclusion or the limitation of employers' responsibility where occurrences are due to unusual and unforeseeable circumstances, beyond the employers' control, or to exceptional events, the consequences of which could not have been avoided despite the exercise of all due care.

Member States need not exercise the option referred to in the first subparagraph.

## Article 6
### General Obligations on Employers

1   Within the context of his responsibilities, the employer shall take the measures necessary for the safety and health protection of workers, including prevention of occupational risks and provision of information and training, as well as provision of the necessary organization and means.

   The employer shall be alert to the need to adjust these measures to take account of changing circumstances and aim to improve existing situations.

2   The employer shall implement the measures referred to in the first subparagraph of paragraph 1 on the basis of the following general principles of prevention:

   (a)   avoiding risks;
   (b)   evaluating the risks which cannot be avoided;
   (c)   combating the risks at source;
   (d)   adapting the work to the individual, especially as regards the design of workplaces, the choice of work equipment and the choice of working and production methods, with a view, in particular, to alleviating monotonous work and work at a predetermined work-rate and to reducing their effect on health;
   (e)   adapting to technical progress;
   (f)   replacing the dangerous by the non-dangerous or the less dangerous;
   (g)   developing a coherent overall prevention policy which covers technology, organization of work, working conditions, social relationships and the influence of factors related to the working environment;
   (h)   giving collective protective measures priority over individual protective measures;
   (i)   giving appropriate instructions to the workers.

3   Without prejudice to the other provisions of this Directive, the employer shall, taking into account the nature of the activities of the enterprise and/or establishment:

   (a)   Evaluate the risks to the safety and health of workers, *inter alia* in the choice of work equipment, the chemical substances or preparations used, and the fitting-out of workplaces.

      Subsequent to this evaluation and as necessary, the preventive measures and the working and production methods implemented by the employer must:

      – assure an improvement in the level of protection afforded to workers with regard to safety and health,

> – be integrated into all the activities of the undertaking and/or establishment and at all hierarchical levels;
>
> (b) where he entrusts tasks to a worker, take into consideration the worker's capabilities as regards health and safety;
>
> (c) ensure that the planning and introduction of new technologies are the subject of consultation with the workers and/or their representatives, as regards the consequences of the choice of equipment, the working conditions and the working environment for the safety and health of workers;
>
> (d) take appropriate steps to ensure that only workers who have received adequate instructions may have access to areas where there is serious and specific danger.

4  Without prejudice to the other provisions of this Directive, where several undertakings share a workplace, the employers shall cooperate in implementing the safety, health and occupational hygiene provisions and, taking into account the nature of the activities, shall coordinate their actions in matters of the protection and prevention of occupational risks, and shall inform one another and their respective workers and/or workers' representatives of these risks.

5  Measures related to safety, hygiene and health at work may in no circumstances involve the workers in financial cost.

*Article 7*
*Protective and Preventive Services*

1  Without prejudice to the obligations referred to in Articles 5 and 6, the employer shall designate one or more workers to carry out activities related to the protection and prevention of occupational risks for the undertaking and/or establishment.

2  Designated workers may not be placed at any disadvantage because of their activities related to the protection and prevention of occupational risks.

   Designated workers shall be allowed adequate time to enable them to fulfil their obligations arising from this Directive.

3  If such protective and preventive measures cannot be organized for lack of competent personnel in the undertaking and/or establishment, the employer shall enlist competent external services or persons.

4  Where the employer enlists such services or persons, he shall inform them of the factors known to affect, or suspected of affecting, the safety and health of the workers and they must have access to the information referred to in Article 10(2).

5  In all cases:

   – the workers designated must have the necessary capabilities and the necessary means,

– the external services or persons consulted must have the necessary aptitudes and the necessary personal and professional means, and

– the workers designated and the external services or persons consulted must be sufficient in number,

to deal with the organization of protective and preventive measures, taking into account the size of the undertaking and/or establishment and/or the hazards to which the workers are exposed and their distribution throughout the entire undertaking and/or establishment.

6    The protection from, and prevention of, the health and safety risks which form the subject of this Article shall be the responsibility of one or more workers, of one service or of separate services whether from inside or outside the undertaking and/or establishment.

The worker(s) and/or agency(ies) must work together whenever necessary.

7    Member States may define, in the light of the nature of the activities and size of the undertakings, the categories of undertakings in which the employer, provided he is competent, may himself take responsibility for the measures referred to in paragraph 1.

8    Member States shall define the necessary capabilities and aptitudes referred to in paragraph 5.

They may determine the sufficient number referred to in paragraph 5.

*Article 8*
*First Aid, Fire Fighting and Evacuation of Workers, Serious and Imminent Danger*

1    The employer shall:

– Take the necessary measures for first aid, fire-fighting and evacuation of workers, adapted to the nature of the activities and the size of the undertaking and/or establishment and taking into account other persons present,

– Arrange any necessary contacts with external services, particularly as regards first aid, emergency medical care, rescue work and fire fighting.

2    Pursuant to paragraph 1, the employer shall *inter alia*, for first aid, fire fighting and the evacuation of workers, designate the workers required to implement such measures.

The number of such workers, their training and the equipment available to them shall be adequate, taking account of the size and/or specific hazards of the undertaking and/or establishment.

3    The employer shall:

(a)    As soon as possible, inform all workers who are, or may be, exposed to serious and imminent danger of the risk involved and of the steps taken or to be taken as regards protection;

(b)    take action and give instructions to enable workers in the event of serious, imminent and unavoidable danger to stop work and/or immediately to leave the workplace and proceed to a place of safety;

(c) save in exceptional cases for reasons duly substantiated, refrain from asking workers to resume work in a working situation where there is still a serious and imminent danger.

4   Workers who, in the event of serious, imminent and unavoidable danger, leave their workstation and/or a dangerous area may not be placed at any disadvantage because of their action and must be protected against any harmful and unjustified consequences, in accordance with national laws and/or practices.

5   The employer shall ensure that all workers are able, in the event of serious and imminent danger to their own safety and/or that of other persons, and where the immediate superior responsible cannot be contacted, to take the appropriate steps in light of their knowledge and the technical means at their disposal, to avoid the consequences of such danger.

Their actions shall not place them at any disadvantage, unless they acted carelessly or there was negligence on their part.

*Article 9*
*Various Obligations on Employers*

1   The employer shall:
(a) be in possession of an assessment of the risks to safety and health at work, including those facing groups of workers exposed to particular risks;
(b) decide on the protective measures to be taken and, if necessary, the protective equipment to be used;
(c) keep a list of occupational accidents resulting in a worker being unfit for work for more than three working days;
(d) draw up, for the responsible authorities and in accordance with national laws and/or practices, reports on occupational accidents suffered by his workers.

2   Member States shall define, in the light of the nature of the activities and size of the undertakings, the obligations to be met by the different categories of undertakings in respect of the drawing-up of the documents provided for in paragraph 1 (a) and (b) and when preparing the documents provided for in paragraph 1 (c) and (d).

*Article 10*
*Worker Information*

1   The employer shall take appropriate measures so that workers and/or their representatives in the undertaking and/or establishment receive, in

accordance with national laws and/or practices which may take account, *inter alia*, of the size of the undertaking and/or establishment, all the necessary information concerning:

(a)  the safety and health risks and protective and preventive measures and activities in respect of both the undertaking and/or establishment in general and each type of workstation and/or job;

(b)  the measures taken pursuant to Article 8(2).

2  The employer shall take appropriate measures so that employers of workers from any outside undertakings and/or establishments engaged in work in his undertaking and/or establishment receive, in accordance with national laws and/or practices, adequate information concerning the points referred to in paragraph 1 (a) and (b) which is to be provided to the workers in question.

3  The employer shall take appropriate measures so that workers with specific functions in protecting the safety and health of workers, or workers' representatives with specific responsibility for the safety and health of workers shall have access, to carry out their functions and in accordance with national laws and/or practices, to:

(a)  the risk assessment and protective measures referred to in Article 9(1) (a) and (b);

(b)  the list and reports referred to in Article 9(1) (c) and (d);

(c)  the information yielded by protective and preventive measures, inspection agencies and bodies responsible for safety and health.

*Article 11*
*Consultation and Participation of Workers*

1  Employers shall consult workers and/or their representatives and allow them to take part in discussions on all questions relating to safety and health at work.

This presupposes:

– the consultation of workers,

– the right of workers and/or their representatives to make proposals,

– balanced participation in accordance with national laws and/or practices.

2  Workers or workers' representatives with specific responsibility for the safety and health of workers shall take part in a balanced way, in accordance with national laws and/or practices, or shall be consulted in advance and in good time by the employer with regard to:

(a)  any measure which may substantially affect safety and health;

(b)  the designation of workers referred in Articles 7(1) and 8(2) and the activities referred to in Article 7(1);

(c)  the information referred to in Articles 9(1) and 10;

(d) the enlistment, where appropriate, of the competent services or persons outside the undertaking and/or establishment, as referred to in Article 7(3);

(e) the planning and organization of the training referred to in Article 12.

3 Workers' representatives with specific responsibility for the safety and health of workers shall have the right to ask the employer to take appropriate measures and to submit proposals to him at that end to mitigate hazards for workers and/or to remove sources of danger.

4 The workers referred to in paragraph 2 and the workers' representatives referred to in paragraphs 2 and 3 may not be placed at a disadvantage because of their respective activities referred to in paragraphs 2 and 3.

5 Employers must allow workers' representatives with specific responsibility for the safety and health of workers adequate time off work, without loss of pay, and provide them with the necessary means to enable such representatives to exercise their rights and functions deriving from this Directive.

6 Workers and/or their representatives are entitled to appeal, in accordance with national law and/or practice, to the authority responsible for safety and health protection at work if they consider that the measures taken and the means employed by the employer are inadequate for the purposes of ensuring safety and health at work.

Workers' representatives must be given the opportunity to submit their observations during the inspection visits by the competent authority.

*Article 12*
*Training of Workers*

1 The employer shall ensure that each worker receives adequate safety and health training, in particular in the form of information and instructions specific to his workstation or job:
– on recruitment,
– in the event of a transfer or a change of job,
– in the event of the introduction of new work equipment or a change in equipment,
– in the event of the introduction of any new technology.
The training shall be:
– adapted to take account of new or changed risks, and
– repeated periodically if necessary.

2 The employer shall ensure that workers from outside undertakings and/or establishments engaged in work in his undertaking and/or establishment have in fact received appropriate instructions regarding health and safety risks during their activities in his undertaking and/or establishment.

3 Workers' representatives with a specific role in protecting the safety and health of workers shall be entitled to appropriate training.

4   The training referred to in paragraphs 1 and 3 may not be at the workers' expense or at that of the workers' representatives.

The training referred to in paragraph 1 must take place during working hours.

The training referred to in paragraph 3 must take place during working hours or in accordance with national practice either within or outside the undertaking and/or the establishment.

# SECTION III
# WORKERS' OBLIGATIONS

*Article 13*

1   It shall be the responsibility of each worker to take care as far as possible of his own safety and health and that of other persons affected by his acts or Commissions at work in accordance with his training and the instructions given by his employer.

2   To this end, workers must in particular, in accordance with their training and the instructions given by their employer:

(a)   make correct use of machinery, apparatus, tools, dangerous substances, transport equipment and other means of production;

(b)   make correct use of the personal protective equipment supplied to them and, after use, return it to its proper place;

(c)   refrain from disconnecting, changing or removing arbitrarily safety devices fitted, e.g.: to machinery, apparatus, tools, plant and buildings, and use such safety devices correctly;

(d)   immediately inform the employer and/or the workers with specific responsibility for the safety and health of workers of any work situation they have reasonable grounds for considering represents a serious and immediate danger to safety and health and of any shortcomings in the protection arrangements;

(e)   cooperate, in accordance with national practice, with the employer and/or workers with specific responsibility for the safety and health of workers, for as long as may be necessary to enable any tasks or requirements imposed by the competent authority to protect the safety and health of workers at work to be carried out;

(f)   cooperate, in accordance with national practice, with the employer and/or workers with specific responsibility for the safety and health of workers, for as long as may be necessary to enable the employer to ensure that the working environment and working conditions are safe and pose no risk to safety and health within their field of activity.

# SECTION IV
# MISCELLANEOUS PROVISIONS

*Article 14*
*Health Surveillance*

1   To ensure that workers receive health surveillance appropriate to the health and safety risks they incur at work, measures shall be introduced in accordance with national law and/or practice.
2   The measures referred to in paragraph 1 shall be such that each worker, if he so wishes, may receive health surveillance at regular intervals.
3   Health surveillance may be provided as part of a national health system.

*Article 15*
*Risk Groups*

Particular sensitive risk groups must be protected against the dangers which specifically affect them.

*Article 16*
*Individual Directives – Amendments – General Scope of this Directive*

1   The Council, acting on a proposal from the Commission based on Article 118a of the Treaty, shall adopt individual Directives, *inter alia*, in the areas listed in the Annex.
2   This Directive and, without prejudice to the procedure referred to in Article 17 concerning technical adjustments, the individual Directives may be amended in accordance with the procedure provided for in Article 118a of the Treaty.
3   The provisions of this Directive shall apply in full to all the areas covered by the individual Directives, without prejudice to more stringent and/or specific provisions contained in these individual Directives.

*Article 17*
*Committee*

1   For the purely technical adjustments to the individual Directives provided for in Article 16(1) to take account of:
    – The adoption of Directives in the field of technical harmonization and standardization, and/or
    – Technical progress, changes in international regulations or specifications, and new findings

the Commission shall be assisted by a committee composed of the representatives of the Member States and chaired by the representative of the Commission.

2   The representative of the Commission shall submit to the committee a draft of the measures to be taken.

The committee shall deliver its opinion on the draft within a time limit which the chairman may lay down according to the urgency of the matter.

The opinion shall be delivered by the majority laid down in Article 148(2) of the Treaty in the case of decisions which the Council is required to adopt on a proposal from the Commission.

The votes of the representatives of the Member States within the committee shall be weighted in the manner set out in that Article. The chairman shall not vote.

3   The Commission shall adopt the measures envisaged if they are in accordance with the opinion of the committee.

If the measures envisaged are not in accordance with the opinion of the committee, or if no opinion is delivered, the Commission shall, without delay, submit to the Council a proposal relating to the measures to be taken. The Council shall act by a qualified majority.

If, on the expiry of three months from the date of the referral to the Council, the Council has not acted, the proposed measures shall be adopted by the Commission.

*Article 18*
*Final Provisions*

1   Member States shall bring into force the laws, Regulations and administrative provisions necessary to comply with their Directive by 31 December 1992.

They shall forthwith inform the Commission thereof.

2   Member States shall communicate to the Commission the texts of the provisions of national law which they have already adopted or adopt in the field covered by this Directive.

3   Member States shall report to the Commission every five years on the practical implementation of the provisions of this Directive, indicating the points of view of employers and workers.

The Commission shall inform the European Parliament, the Council, the Economic and Social Committee and the Advisory Committee on Safety, Hygiene and Health Protection at Work.

4   The Commission shall submit periodically to the European Parliament, the Council and the Economic and Social Committee a report on the implementation of this Directive, taking in account paragraphs 1 to 3.

*Article 19*

This Directive is addressed to the Member States.

Done at Luxembourg, 12 June 1989.

*For the Council*
*The President*
*M. CHAVES GONZALES*

---

*ANNEX*
*List of Areas Referred to in Article 16(1)*

– Workplaces
– Work equipment
– Personal protective equipment
– Work with visual display units
– Handling of heavy loads involving risk of back injury
– Temporary or mobile work sites
– Fisheries and agriculture

# Appendix II: Planning supervisor's pre-tender health and safety plan

**CONTENTS**

## 1. Introduction

### 1.1 Generally

All prospective principal contractors tendering for this contract will receive this Health and Safety Plan. The purpose is to highlight the main health and safety issues in connection with the construction work on the project and to form a basis for tenderers to explain their proposals for managing the problems.

The Principal Contractor will develop this Health and Safety Plan, in particular taking reasonable steps to ensure cooperation between all contractors to achieve compliance with the rules and recommendations of the plan.

### 1.2 Method statements

Detailed method statements, produced by the Principal Contractor before work starts, are essential for safe working and is a requirement of Section 2(2) of the Health and Safety at Work, etc. Act. They should identify problems and their solutions and form a reference for the site supervision. The method statements should be easy to understand, should be agreed by and known to all levels of management and supervision and should include such matters as outlined in this document.

### 1.3 Health and Safety File

A Health and Safety File for the site is available for inspection by appointment during normal office hours.

A copy of the Health and Safety Plan will be handed to the Principal Contractor for use during the Works and for developing for final handover to the Client.

### 1.4 Questionnaire

Each tenderer is required to complete a questionnaire outlining their experiences of health and safety on site. This questionnaire is provided as a separate document. The questionnaire is to be completed and returned to the address below no later than one week before the date for return of tenders.

## 2. Nature of the project

### 2.1 The Client

### 2.2 Client's agent

### 2.3 Location

### 2.4 Nature of the Works

#### 2.4.1 Generally

Hard and soft landscape works on remediated land together with works to the riverside.

### 2.4.2   General Landscape Works

- Hard and soft landscaped areas
- Footpaths
- Planting trees and shrubs
- Boundary railings
- Site furniture
- Creation of plaza including feature screen
- Creation of landscaped mounds

### 2.4.3   Civil Engineering Work

- Foundations and Sub-bases
- Riverside walkways
- Walls and piers
- Work to retaining wall
- Land drainage and other drainage
- Head walls

### 2.4.4   Play area and equipment

- Construction of bark pits
- Construction of drainage sumps
- Landscaping
- Play equipment and site furniture
- Fencing

### 2.4.5   Street Lighting

- Lighting installations to new and existing pedestrian routes

## 2.5   Timescale
Contact period

The Contractor is to refer to the Preliminaries/General Conditions for details of Contract dates for Sectional Completion.

## 3.   The existing environment

### 3.1   Site description
The site occupies a flat, low lying area situated between the river and the canal to the west. The river and canal are joined to the north and south of

the site, so forming an island. The site occupies a total area of approximately 15 hectares.

A pedestrian route runs across the site from east to west. The Contractor is to allow for maintaining this footpath and for ensuring the safety of all pedestrians using it.

The site is currently secured by a 2.4 metre high hoarding around the perimeter of the site and between the park and the proposed residential area. There is no hoarding between the park and the proposed business park.

### 3.2 Existing land, ground conditions and nature of soil

Former land uses included warehousing, metal recycling or 'scrap yards', railway goods and maintenance yard and a power station. Ground contaminations were identified arising out of these previous site uses. All such materials have since been removed by specialists as part of a separate reclamation contract.

The base of reclamation excavations for the proposed park area was generally 58.00 AOD. At the entrance the base of reclamation excavations was lowered to 56.00 AOD. The 'as constructed' reclamation finished levels provided by the reclamation contractor are presented on drawings in the Tender Documents.

Full details of the reclamation contract are included in the Health and Safety File.

### 3.3 Ground water level

The ground water level on the site is described in the Trade Preambles for Excavations and Hard Landscape Work.

### 3.4 Existing structures

Prior to reclamation, all buildings, except the 'Machine House', were demolished. The 'Machine House' comprises a brick built structure and has two main chambers, accessed from the north and south side of the building respectively. It also has a tower which was believed to house a steam driven hydraulic pump.

While work to this structure does not form part of this contract, work will be going on around and adjacent to it and the Principal Contractor should make themselves aware of its condition.

### 3.5 Surrounding land uses

There are buildings close by, therefore, the Principal Contractor can expect busy pedestrian traffic.

This pedestrian route will remain open at all times. This is a temporary foot-path which will be relocated under this contract.

The cycle path, running north to south along the western side of the site, is also to remain open at all times. Part of this path is to be resurfaced and part is to be re-routed under this contract.

The cycle path, running north to south along the western side of the site, is also to remain open at all times. Part of this path is to be resurfaced and part is to be re-routed under this contract.

The Principal Contractor shall adopt suitable working methods to ensure the safety of all members of the public on these pedestrian and cycle routes.

### 3.6 Proposed adjacent development
To the north and south of the Park are areas of proposed development.

These proposed developments comprise a business park to the north and a residential area to the south. The developer of the residential area is currently on site and the Contractor is to assume that the developers for both that area and the science park will be on site during the course of the Park contract.

The Principal Contractor is to take into account the possibility of:
– Overhead working on the adjacent site
– The erection of building components using cranes on adjacent sites.

The Principal Contractor is to detail any temporary fencing and signage to be used in advance of the construction of permanent boundaries.

### 3.7 Restricted access
There is restricted access to parts of the site due to neighbouring properties. Care is to be taken while working in these areas and the entrances to proper-ties must remain open at all times.

### 3.8 Existing services
The extent of known site services information is indicated on drawings and also on various drawings in the Health and Safety File. Where positions of existing services are indicated on the drawings they are for guidance only; it is the responsibility of the Contractor to obtain details and exact positions from relevant service authorities.

Existing services in the public footpaths along the route across the site will actionill remain live. Alterations to the street lighting installations will form part of the Works.

Considering the depth of remediations it is not anticipated that any other underground services will be encountered in normal excavations. However, the Contractor is to proceed with caution during all excavations, particularly during excavations below the remediation level, and is to establish formal procedures to ensure that any unidentified services located during the works are carefully checked to determine if they are live or contain any hazardous material or substances. All details must be recorded. The Services Engineer and the Planning Supervisor are to be provided with copies of the records setting out the nature and location of all such services prior to the agreement of a course of action.

### 3.9   Traffic systems and restrictions

The Principal Contractor is to determine and apply all Local Authority and Police Regulations.

### 3.10   Health hazards

#### 3.10.1   Asbestos

Asbestos may be present around the 'machine house'. The removal of any material containing asbestos must be carried out strictly in accordance with the following statutory measures:

a)   Health and Safety at Work etc. Act, 1974
b)   Control of Pollution Act, 1974
c)   Control of pollution (Special Waste) Regulations 1980
d)   Asbestos (Licensing) Regulations, 1983
e)   Control of Asbestos at Work Regulations 1987
f)   Any other legislation that may be applicable.

The Principal Contractors and its Subcontractors are responsible for the safety, health and welfare of their 'own employees'.

Attention is also drawn in particular to the requirements of the 1983 Regulations:

(a)   for all operations to be carried out by a licensed contractor
(b)   for 28 days notice, or such shorter period as may be agreed, to be given to the relevant Enforcement Authority, whatever type of asbestos is involved. If in doubt about the identity of the relevant Enforcement Authority the Contractor should contact the HSE.

#### 3.10.2   Hazardous material generally

Hazardous material present within the site area, particularly on the embankments of the river, may include broken glass and used syringes among other domestic debris. Suitable precautions should be taken to prevent injury to the workforce or members of the public and all broken glass and dangerous objects removed from site.

## 4. Existing drawings and documents

### 4.1 Health and Safety File
A Health and Safety File for the site is available for inspection by appointment during normal office hours.

A copy of the Health and Safety Plan will be handed to the Principal Contractor for use during the Works and for developing for final handover to the Client.

### 4.2 Drawings included in the Health and Safety File
List as required.

### 4.3 Other documents included in the Health and Safety File
List as required.

### 4.4 Existing Drawings included as part of the Tender Documents
List as required.

## 5. The design

The following principal hazards or work sequences so far identified cannot be avoided and will be a risk to health and safety of construction workers and/ or the occupants of, or visitors to, the site.

It will be the responsibility of the Principal Contractor to detail their proposals for managing these problems. These details/method statements will be incorporated into the health and safety plan prior to the work commencing on site.

### 5.1 Working over water
Care is to be taken while working adjacent to the water courses. The Contractor should carefully consider the method of working and material handling procedures. Construction operatives must be suitably trained and qualified. Proper equipment such as scaffolds, platforms, safety nets, safety belts, harnesses and lanyards should be considered. Suitable emergency procedures must be used such as the use of lifejackets, buoyancy aids and rescue lines,

together with the information and training necessary to use the equipment. The construction operatives must be capable of responding effectively to an emergency.

The main hazards associated with working over water are:

– falls from heights
– drowning
– disease.

### 5.1.1   Falls from heights

The Contractor is to submit details preventing personnel and debris from falling into the water courses.

### 5.1.2   Drowning

All personnel working in the vicinity of the watercourses must wear buoyancy aids.

In the event of a person falling into the water two things are of paramount importance:

– the person must be kept afloat
– location and rescue must be achieved as quickly as possible.

The Contractor is to submit details for achieving these aims including details of buoyancy aids and rescue equipment, if any, to be employed and details of rescue procedures. In addition, details are required of training given to personnel in order to effect such rescues.

### 5.1.3   Disease

With work being carried out adjacent to the waterways, the Contractor's attention is drawn to the possible presence of vermin and all the associated health hazards.

For further information the Contractor is recommended to read the CITB Construction Site Safety Note 30 'Working Over Water'.

A document titled 'Special Requirements in Relation to the National Rivers Authority' is included as Appendix B. The Contractor is to comply with all such requirements.

The installation of gabions along the edge of a section of the river is required.

As well as all the risks associated with working over water as identified in item 5.1 the installation of the gabions poses other problems, including:

– Proximity to the old railway bridge
– Limited headroom.

The Principal Contractor is to submit a method statement demonstrating how the installation of the gabions will be achieved.

### 5.2 Excavations below level of remediation
Excavations below the base of the reclamation excavations will generate material not tested as part of the reclamation.

### 5.3 General excavations greater than 1.2m deep
Most excavations will be in made up ground. The Contractor's method statements must identify proposals for:

– The prevention of collapse
– The protection of edges of excavations to prevent falls of personnel, plant or materials into the trench.

### 5.4 Drainage work
The main problems associated with the proposed drainage work will be:

– Deep excavations (including work below remediation level)
– Work in confined spaces.

### 5.5 Connections to existing services
The Contractor's proposals are required for safe working methods for connections to the following live services:

– Electricity
– Water
– Sewerage.

### 5.6 Work on steep slopes
Work on steep slopes includes landscape works to mounds and the clearing of rubbish and debris to the banks of the river. The main hazards associated with working on steep slopes include:

– Angle of slope
– Slippery surfaces
– Hidden holes or obstacles

– Visibility

– Stability of machinery on bank

– Position of body while carrying out the work.

Care is to be taken when carrying out these works and the work is to be undertaken by suitably trained and qualified operatives wearing suitable protective clothing. Appropriate plant is to be used.

## 5.7 Street lighting installations

The work includes the disconnection and removal of existing street lighting columns along the pedestrian routes and the installation of new cabling and columns. The existing cabling is to remain in the ground; the Contractor is to ensure that it is disconnected and made safe.

The risks associated with the street lighting installations include:

### 5.7.1 Electrocution and fire

These risks may be avoided by compliance with:

1   Electricity at Work Regulations
2   IEE Wiring Regulations
3   Health and Safety at Work, etc. Act
4   Managing Health and Safety on Construction Sites
5   Electrical Specification.

No installation is to be energised until installation is complete, inspected and tested and a compliant 'Electrical Installation Certificate' is received.

All trades are to be made aware of the hazards and adequate protection is to be provided to safeguard staff and to prevent unauthorized entry.

### 5.7.2 Risks to the public

Much of the work associated with the street lighting installations will be carried out on existing pedestrian routes which are to remain open at all times. Suitable signs are to be erected warning members of the public of the work being undertaken.

### 5.7.3 Falling/tripping

All open excavations are to be adequately protected and are to be backfilled as soon as possible. All materials are to be stored away when not required.

### 5.7.4 General site activities/Contractor's staff

Ensure the provision of personal protective equipment.

### 5.7.5 Manual handling of heavy equipment

Ensure compliance with Manual Handling Operations Regulations 1992.

## 5.8 Feature screen to plaza housing access features

A six metre high stainless steel feature screen is to be erected in the entrance plaza. The main hazards associated with this aspect of the work include:

### 5.8.1 Manual handling of heavy equipment

Ensure compliance with Manual Handling Operations Regulations 1992.

### 5.8.2 On-site welding operations

Ensure safe working methods.

## 5.9 General manual handling

The Contractor's proposals are required for the safe manual handling of such items as:

– Railings and gates
– Components of octagonal feature piers
– Pier caps and wall copings
– Lighting columns
– Bollards
– Inspection chamber covers
– Rootballed trees
– Seats and other street furniture.

## 5.10 Work to existing trees

The Principal Contractor is to prepare a method statement for safe working.

## 5.11 Maintenance of soft landscaped areas

At various stages during the works and following practical completion, the Contractor will be required to carry out maintenance work to soft landscaped areas. Problems arising out of such work will include:

– Conflict with members of the public
– The use of vehicles in areas used by the public
– The use of chemicals in areas used by the public.

Particular care is to be used during such operations and the work is to be carried out by suitably trained and qualified operatives in accordance with all current Regulations and enactments.

## 6. Construction materials

### 6.1 Construction materials generally

The following potentially hazardous construction materials and substances are required by the design.

### 6.1.1 Examples of substances harmful by inhalation

– Welding fumes
– Hardwood dust
– Cement dust
– Isocyanates (paints, varnishes, adhesives)
– Solvents (paints, strippers, mastics, glues, surface coatings).

### 6.1.2 Examples of substances hazardous in contact with skin and mucous membranes

– Bitumen
– Brick, concrete, stone dust
– Cement
– Paints, varnishes, stains
– Certain epoxy resins
– Chromates (in primer paints, cement)
– Petrol, white spirit, thinners
– Acids
– Alkalis.

The Principal Contractor is required to take appropriate measures to control the risks created by these hazards, to detail these proposed control measures in the Health and Safety Plan and to prepare detailed method statements for managing these aspects of the works.

Other common materials and substances used during construction will also present health and/or safety hazards, requiring the contractor to carry out COSHH or other risk assessments and to introduce control measures. These are deemed to be within the normal experience of a competent contractor and have not, therefore, been listed here.

### 6.2 Hot bituminous material

Care is required while working with hot bituminous materials. Suitable protective clothing and footwear must be worn.

## 7. Site-wide elements

### 7.1 Site access and egress points
The main vehicular access to the site will be at the west entrance. A secondary site access is available to the north of the site.

### 7.2 Movement of site traffic
The Principal Contractor must liaise with other contractors on adjacent sites regarding the movement of vehicles around the site and around the immediate area.

It will be the responsibility of the principal contractor to detail their proposals for managing the movement of all site traffic, including their own plant, those of subcontractors, delivery vehicles and the like, around the site and around the immediate area. The effects of site traffic on the local environment and upon people living and working in the area are to be kept to a minimum.

These details/method statements will be incorporated into the health and safety plan prior to the work commencing on site.

### 7.3 Location of temporary site accommodation
The Principal Contractor's details are required for the proposed location of temporary site accommodation and of the site compound. The proposed location of such accommodation is to be indicated on a drawing which will form part of the construction phase health and safety plan.

### 7.4 Location of unloading, layout and storage areas
Vehicles to be loaded/unloaded under supervision by a competent person.

No vehicles will be allowed to wait on main roads.

A method statement is required detailing the procedure for loading/offloading materials and for maintaining stability and tidiness of stacks and containers.

### 7.5 Traffic/pedestrian routes
All traffic and pedestrian routes, except where closed to the public as indicated on the drawings, must remain open at all times and must remain free from spoil and debris. Vehicle access has to be maintained for the public and for emergency services at all times. All measures are to be taken to minimize conflicts between the public and site traffic.

All routing of site traffic, signing and barriers are to be in accordance with safe site practice and Local Authority requirements. A stopping up order will be required for any essential temporary closures.

The Contractor is to prepare method statements for phasing and safe working.

### 7.6 Site security

The site is currently secured by a 2.4 metre high hoarding around the perimeter of the site. The Contractor is to allow for regular inspection and maintenance of all such hoardings during the Works.

At various stages of the Works several lengths of the security hoarding will need to be dismantled and re-erected to allow certain aspects of the work to be carried out. The permanent removal of this hoarding will form part of these Works; they are to be dismantled:

a) on completion of contract works, or
b) on completion of permanent boundaries, or
c) to facilitate works, in which event alternative measures are to be taken to ensure public safety.

### 7.7 Coordination with work to be executed by others

The following work is to be carried out by other contractors:

#### 7.7.1 Installation of play equipment

#### 7.7.2 Outfall to housing development

The Contractor is to obtain from, and arrange with, the other contractors, details of times of commencement of work and delivery of plant and materials, phased to suit the Contractor's programme and must allow all reasonable access.

The Contractor is to be the Principal Contractor for these aspects of the work. Safe methods of work are to be agreed between all contractors. All personnel are to be made aware of general health and safety requirements demanded elsewhere in the health and safety plan.

### 7.8 Site welding

Site welding operations will be required to the components of the feature screen to the plaza and to other aspects of the Works. All precautions are to be taken to prevent injury not only to the personnel carrying out the works but also to personnel not actually engaged in the operation and also to members of the public.

## 8.  Overlap with client's undertaking

### 8.1 Buildings in occupation

It must be acknowledged and accepted by the Principal Contractor that adjacent properties will remain occupied throughout the duration of the contract.

Specific provision must be made for the protection of the public especially while contractor's plant is moving around the area.

All existing services must be maintained.

The Contractor's proposals are required for managing these problems.

## 9. Site rules

### 9.1 Generally
The Principal Contractor is reminded of the high public profile which the client enjoys and shall therefore ensure that no site activity reflects adversely on the client.

The Principal Contractor's proposals for complying with these site rules are to be set out in the construction phase health and safety plan and must include provisions for:

### 9.2 Conformity with all statutory requirements
Including:

– The Construction (Health, Safety and Welfare) Regulations 1996
– Electricity at Work Regulations
– PPE at Work Regulations
– COSHH Regulations
– Noise at Work Regulations.

### 9.3 Booking in/out
The Contractor's proposals are required for ascertaining what visitors are present on site and ensuring they are aware of the site risks.

### 9.4 Sub-soil stockpile
The provision of a sub-soil stockpile may mean double handling of material which increases the possibility of site traffic conflicts and may also impose restrictions on site working.

Work is to be carried out by suitably trained and qualified operatives. The Contractor is to allow for all effects in programming works and is to prepare a method statement for safe working, identifying any temporary stockpile areas required.

### 9.5 Topsoil storage area
The risks involved include site traffic conflicts and conflict with members of the public.

Work is to be carried out by suitably trained and qualified operatives.

The Contractor is to prepare a method statement for safe working, identifying proposed phasing of the works and haulage routes.

### 9.6   Control of movements

The Contractor's proposals are required for controlling the movements around the site of all of occasional personnel.

### 9.7   'Absence' procedure

The Contractor's proposals are required for appointing a person responsible for ensuring the management of health and safety on site in the absence of the named person in the health and safety plan.

The proposals are to cover short absences of less than one day and for planned and unplanned absences for one full day or more.

### 9.8   Unforeseen eventualities

A formal reporting system for unforeseen eventualities is required.

### 9.9   Fires on site

Fires will not be permitted on the site.

### 9.10   Site security

The Contractor's proposals are required for ensuring that the public and the like are kept off site especially outside normal working hours. Detailed requirements are set out in the Bills of Quantities. The Contractor is to state how this aspect of security is to be managed.

The Contractor is required to detail proposals for dealing with regular inspections and repair of damaged or vandalised fencing.

### 9.11   Access around the site

The Contractor's proposals are required for identifying walkways and routes and ensuring that they remain clear and unrestricted at all times. Proposals should cover for all operatives and visitors for all stages of construction.

The Contractor's proposals are required for clearing away all debris and leaving the site reasonably tidy at the end of each working day. Rubbish and scattered building materials will not be allowed to accumulate.

### 9.12   Wheel wash facility

The Contractor is to set up a wheel wash facility to ensure that all roads and footpaths remain free from spoil and debris at all times.

### 9.13 Provision of welfare facilities

Provide all necessary welfare facilities including sanitary conveniences and washing facilities to comply with Regulation 22 of The Construction (Health, Safety and Welfare) Regulations.

### 9.14 First aid

Adequate first aid equipment is to be made available and the Contractor must ensure the presence on site of a person trained in its use at all times.

Specific proposals are required for the treatment of the following:

a) the risk of drowning and associated hazards,
b) the risk of electrocution especially where electrically operated tools are used in the vicinity of water,
c) the recognition and preliminary treatment of leptosperosis.

### 9.15 Personnel on site

Considering the remoteness of certain areas of the site, no personnel are to be allowed to work alone in such locations. This particularly applies where work is being undertaken in close proximity to the watercourses.

### 9.16 Communication around the site

The use of mobile phones and/or walkie-talkies is recommended for communication between remote areas and for use in emergencies.

### 9.17 Control of noise

The Contractor's proposals are required for ensuring that limits described in the Preliminaries are not exceeded.

The Contractor's proposals are also required for ascertaining noise levels on site at regular intervals and when requested by the Client's representative.

Comply generally with BS 5228 and all relevant legislation.

Fit all compressors, percussion tools and vehicles with effective silencers of a type recommended by manufacturers of the equipment.

Other possible hazards to be taken into consideration include vibration and the effect that these have on the workforce as well as on the public. Ear protection should be considered during high-level noise operations.

### 9.18 Control of pollution

The Contractor's proposals are required for ensuring that the site is kept clean and for ensuring the control of dust.

The Contractor's proposals are required for ensuring that debris does not fall into the water courses.

### 9.19   Training

The Contractor's proposals are required for induction, ascertaining general needs and specific training for identified risks.

### 9.20   Competence of site personnel

The Contractor's proposals are required for vetting and for recording the competence of all personnel and other Contractors' personnel for all stages of construction and management of the site.

## 10.   Continuing liaison

### 10.1   Generally

Procedures for dealing with unforeseen events during the project which result in substantial design changes and which might affect resources are as follows:

> **10.1.1**   In the event of any unforeseen circumstance, the Planning Supervisor is to be informed immediately by the Principal Contractor.

> **10.1.2**   Details of the health and safety issues arising from any unforeseen occurrence are to be submitted to the Planning Supervisor as soon as is possible after the event.

> **10.1.3**   In the event that any redesign is required, for whatever reason, the health and safety implications are to be submitted for consideration and acceptance by the Planning Supervisor in due time before execution.

### 10.2   Recording health and safety matters

The Principal Contractor's proposals are required for recording and resolving health and safety matters realized by contractor's personnel, other contractors, Client's representatives, members of the public and any other affected persons.

## 11.   Tender stage method statements

Tender Stage Method Statements must be submitted with the tender *outlining* safe working methods for the following:

– Excavation below and beyond the remediation level.
– Work in proximity to the water courses.
– Work on or adjacent to public rights of way.
– Coordination with work to be executed by other contractors on adjacent developments.
– Haulage and site traffic routes.
– Stockpile areas.
– Coordination with work to be executed by others (such as play equipment, outfall from housing development).

The method statements must describe safe working methods for these items and must also describe how and when the Contractor proposes to undertake the work.

The Principal Contractor may, at his discretion and at the same time, submit method statements for other parts of the Works.

# Appendix III: Principal contractor's construction health and safety plan

## Introduction

This Health and Safety Plan has been produced to detail the Contractor's approach to the Control of Safety on the Project. The layout of the plan is in accordance with the requirements of the Construction (Design and Management) Regulations 1994.

The Health and Safety Plan has been developed from the Pre-tender Health and Safety Plan prepared by the Planning Supervisor.

### Relevant Legislation

- Workplace (Health, Safety and Welfare) Regulations 1992
- Management of Health and Safety at Work Regulations 1992 amended 1994
- Personal Protective Equipment at Work Regulations 1992
- Manual Handling Operations Regulations 1992
- Provision and Use of Work Equipment Regulations 1992
- Health and Safety (Display Screen Equipment) Regulations 1992
- Control of Substances Hazardous to Health Regulations 1988 amended 1991
- Health and Safety (First Aid) Regulations 1981
- Reporting of Injuries, Diseases and Dangerous Occurrences Regulations 1995
- Health and Safety (Information to Employees) Regulations 1989
- Road Traffic (Carriage of Dangerous Substances in Packages etc.) Regulations 1992
- Road Traffic (Training of Drivers of Vehicles Carrying Dangerous Goods) Regulations 1992
- Noise at Work Regulations 1989
- Construction (Lifting Operations) Regulations 1961
- Construction (Head Protection) Regulations 1989
- Construction (Design and Management) Regulations 1994
- Construction (Health, Safety and Welfare) Regulations 1996
- Environmental Protection Act 1990

- Waste Management Licensing Regulations 1994 amended 1995
- Control of Pollution (Amendment) Act 1989
- Controlled Waste (Registration of Carriers and Seizure of Vehicles) Regulations 1991
- Water Resources Act 1991
- Water Industry Act 1991

## Project policy

It is the policy of the organization that we will at all times conduct our activities so that we will, as far as possible, carry out the works so that they exceed the minimum requirements of the relevant legislation.

We will ensure that optimum priority is given to the health and safety of employees and all other persons affected directly by our activities.

We will ensure that every effort is made to minimize any negative impact upon the environment which any activities may have.

Every employee of the organization has direct responsibility for the health and safety of all persons involved by our actions.

Good safety practices will be a primary concern on this project.

## Description of works

Civil engineering, external and landscape works, including earth moving and fencing to approaches and periphery of the development.

These include:

- earthmoving and topsoiling
- planting trees and shrubs
- grass seeding
- fencing works
- maintenance works.

These are detailed in accompanying documentation and should be read in conjunction with this document.

## Project information

Project Name:
Site Address:
Contact:
Site Telephone No:

Client:
Contact:
Telephone:
Fax:

Planning Supervisor:
Contact:
Telephone:
Fax:

Principal Contractor:
Contact:
Telephone:
Fax:

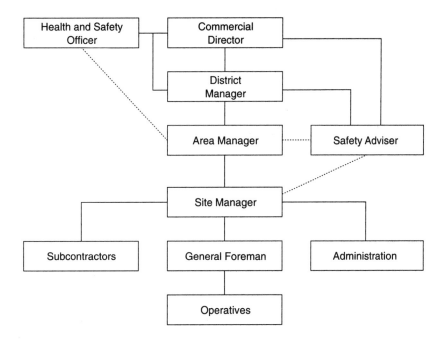

**A1** Organizational
responsibility chart

## Emergency contacts

Local Fire:            for advice Tel:

for emergency 999

Environmental Health Officer:    Tel:

Accident and Emergency Hospital:    Tel:

Area HSE Office:    Tel:

Police:    Tel:

## Risk assessments/safe working procedures

### Hazard identification

Generic hazards identified for which standard risk assessment records and safe working procedures have been prepared and are provided for all personnel on site. These are as follows:

- Unloading/loading by machine
- Grass cutting
- Use of tractors and attachments
- Loading and securing power machinery onto transport
- Pedestrian operated power equipment
- Excavation by machinery
- Transport of materials and equipment
- Application of chemicals
- Use of non-powered hand tools
- Use of powered hand tools
- Working above ground level
- Carrying out operations near roads or water bodies
- Maintenance of vehicles and machinery
- Fence wire tensioning
- Application of protective treatments
- Excavations by hand
- Manual handling
- Hand weeding, litter collection
- Handling cement by hand
- High/unusual risk activities arising from this specific site
- Site/works specific (identified by Planning Supervisor)
- Working on slopes
- Lighting and cable installation
- Risk to pedestrians in work associated with pavements
- Lifting and manoeuvring heavy/awkward objects such as pergola, inscription panel and root balled trees.

**COSHH**

A separate COSHH file will be kept on site for all materials/products incorporated and/or used on the contract. This will be incorporated into the Health and Safety File on completion of the works.

## Management of works on site

### Welfare, first aid and accidents

### Site Accommodation

The site will be kept secure for the storage of equipment and for the provision of appropriate welfare facilities for all employees on site. Secure storage will be provided for work clothing and PPE for all those for whom it is required. Toilet facilities are provided within the compound, adequate for the numbers of personnel likely to be on site at any one time.

Within the compound there will be accommodation for the taking of meals which are brought to site. There will be a supply of drinking water and sufficient water provided for the maximum numbers of people who are expected to be on site.

A qualified first aider is employed and will be based on site with an appropriate first aid kit for the anticipated number of personnel likely to be on site at any one time. The first aider can be contacted on Tel:

A Health and Safety Policy statement, all statutory notices and employees liability insurance certificate will be displayed within the accommodation. Abstracts from the relevant legislation may, if appropriate, also be displayed.

### Accident reporting procedures

An Accident Book will be kept on site at all times. The procedure followed will be as described in the 'Incident Reporting Procedure'. This meets the requirements set out by RIDDOR.

## Management and interface with contractors

All contractors are to complete the Pre-qualification Questionnaire for contractors to assess suitability and competence.

Prior to work starting on site a pre-contract meeting is to be held covering the following topics:

- Induction of employees
- Risk assessments and safe working procedures
- Control of hazardous substances
- Personal protective clothing/equipment
- Operations and use of plant and machinery
- Welfare and first aid arrangements.

### Site Induction

All operatives and management will undergo induction training prior to being allowed for work on site. A standard checklist will be used for:

- site details, i.e. address, telephone and fax numbers
- safety responsibilities to safety management team
- site security arrangements and working hours
- location of first aid equipment in site office
- name of trained first aider
- canteen/messroom arrangements
- toilet and washing facilities
- smoking policy
- fire procedures
  - escape routes
  - exits
  - assembly points
  - location and type of fire extinguishers
- disciplinary procedures
- risk assessment/safe working procedures applicable to employees
- site safety rules.

### Mechanical Plant

Mechanical plant will only be used by operatives trained in its use. Hired plant must:

- arrive on site with confirmation that it has been maintained to the required standard
- must be maintained in good working order and free from defects while on site.

Operators with hired plant must be competent and able to demonstrate their training background with recognized certification.

Copies of all certificates will be kept in the site office. Equipment requiring certification will not be permitted on site unless it is accompanied by a current certificate.

## Site traffic/site restrictions

A number of roads around the site are available to other users and therefore, in order to reduce the risk of injury, certain precautions will be observed as follows:

- All site personnel and visitors will wear high visibility vests on site at all times.
- All lorries will be controlled by a banksman while manoeuvring on site.
- An on-site speed limit of 15 mph will be instructed and enforced.
- All drivers of site vehicles or plant will be in possession of a current full driving licence.

## Excavations

It is not anticipated that any excavations in excess of 2.4m depth will be required during the course of this project. If any are required then appropriate precautions will be taken and further risk assessments and safe working procedures prepared.

It is envisaged that all excavation below levels of remediation will involve the use of machinery. The Safe Working Procedure applicable will apply with the following additional precautions:

- All operators, banksman and workmen involved in excavation work to be briefed on the implications as described in the Planning Supervisor's Report.
- All substances other than standard stone/soil fill material are to be reported to the Site Supervisor immediately they are revealed. Excavation should cease until further notice.

Special attention is to be given to all liquids, metals, containers, coloured glazing of the soil and odours emanating from the excavation.

## Existing services

Where there is a likelihood of encountering any services the appropriate action will be taken.

Hazardous activities are as follows:

- Excavation by machine
- Excavation by hand
- Installation of electrical services.

Precautions necessary:

- See Safe Working Procedures
- Electrical contractor to prepare risk assessment and safe working procedures
- Prior to excavation all services marked out
- Trial holes to be dug prior to deep excavations in the vicinity of the services
- Cable detection equipment and banksman to be used in all circumstances.

## Confined spaces

Any operation which requires entry into a confined space will require a Permit To Work.

Prior to the work being carried out, a comprehensive Hazard Identification and Risk Assessment will be carried out in consultation with our Safety Adviser.

Any necessary escape equipment which is identified as being required will be obtained prior to the work commencing.

The Permit to Work will be issued by the Project Manager and will be kept in the Site Office.

During the operation of a Permit to Work a sign will be displayed which forbids entry to unauthorized personnel.

## Noise

It is not anticipated that noise levels will exceed unobtrusive levels but random assessments will be carried out and if any problem is identified action will be taken as necessary.

## Monitoring procedures

The site manager's responsibilities include inspection to ensure that health and safety standards on site are maintained. A site safety inspection report form is to be completed at least weekly. Enforcement notices can be served by the site manager either:

- an immediate action notice requiring the contractor to confirm the remedial action taken, or
- a stop notice requiring the contractor to cease the activity for repeated contravention or for a serious breach of health and safety.

## Safety plan review

This safety plan will be reviewed regularly to ensure that the information is up-to-date.

Items of on-going management are identification of further hazards, necessity for additional procedures and alterations to training needs.

*Note:*
This document should be read in conjunction with the master site-layout plan and temporary services/works layout plans.

# Select bibliography

British Standards Institution (BSI) (1997) *BS8750: Guide to Occupational Health and Safety Management Systems*, BSI.

Croner (1994) *Croner's Management of Construction Safety*, Croner Publications Ltd.

European Commission (EC) (1989) *Directive 89/391/EEC on the introduction of measures to encourage improvements in the health and safety of workers at work*, HMSO, London.

Health and Safety Commission (HSC) (1996) *Health and Safety Statistics 1995/96*, HMSO, London.

Health and Safety Executive (HSE) (1961) *The Factories Act 1961*, HMSO, London.

Health and Safety Executive (HSE) (1961) *The Construction (General Provisions) Regulations 1961*, HMSO, London.

Health and Safety Executive (HSE) (1961) *The Construction (Lifting Operations) Regulations 1961*, HMSO, London.

Health and Safety Executive (HSE) (1966) *The Construction (Working Places) Regulations 1966*, HMSO, London.

Health and Safety Executive (HSE) (1966) *The Construction (Health and Welfare) Regulations 1966*, HMSO, London.

Health and Safety Executive (HSE) (1974) *The Health and Safety at Work, etc. Act 1974*, HMSO, London.

Health and Safety Executive (HSE) (1988) *Blackspot Construction: A study of five years fatal accidents in the building and civil engineering industries*, HMSO, London.

Health and Safety Executive (HSE) (1992) *The Workplace (Health, Safety and Welfare) Regulations 1992*, HMSO, London.

Health and Safety Executive (HSE) (1992) *The Provision and Use of Work Equipment Regulations 1992*, HMSO, London.

Health and Safety Executive (HSE) (1992) *The Personal Protective Equipment at Work Regulations 1992*, HMSO, London.

Health and Safety Executive (HSE) (1992) *The Manual Handling Operations Regulations 1992*, HMSO, London.

Health and Safety Executive (HSE) (1992) *The Health and Safety (Display Screen Equipment) Regulations 1992*, HMSO, London.

Health and Safety Executive (HSE) (1994) *The Construction (Design and Management) Regulations 1994*, HMSO, London.

Health and Safety Executive (HSE) (1994) *CDM Regulations: How the Regulations Affect You*, HMSO, London.

Health and Safety Executive (HSE) (1995) *The Reporting of Injuries, Diseases and Dangerous Occurrences Regulations 1995*, HMSO, London.

Health and Safety Executive (HSE) (1995) *A Guide to Managing Health and Safety in Construction*, HMSO, London.

Health and Safety Executive (HSE) (1995) *Managing Construction for Health and Safety*, HMSO, London.

Health and Safety Executive (HSE) (1999) *The Management of Health and Safety at Work Regulations 1999*, HMSO, London.

Royal Institute of British Architects (RIBA) (1992) *Standard Form of Agreement for the Appointment of an Architect (SFA/92)*, RIBA.

Royal Institute of British Architects (RIBA) (1995) *Form of Appointment as Planning Supervisor (PS/45)*, RIBA.

Royal Institute of British Architects (RIBA) (1995) *Conditions of Engagement for the Appointment of an Architect (CE/95)*, RIBA.

# Index